ANNALES ACADEMIÆ SCIENTIARUM FENNICÆ

MATHEMATICA

DISSERTATIONES

155

MEROMORPHIC SOLUTIONS OF DIFFERENCE PAINLEVÉ EQUATIONS

ONNI RONKAINEN

University of Eastern Finland, Department of Physics and Mathematics

To be presented, with the permission of the Faculty of Science and Forestry of the University of Eastern Finland, for public criticism in Auditorium M100 of the Joensuu campus, on April 8th, 2010, at 12 o'clock noon.

HELSINKI 2010
SUOMALAINEN TIEDEAKATEMIA

ISSN-L 1239-6303
ISSN 1239-6303 (Print)
ISSN 1798-2375 (Online)
ISBN 978-951-41-1046-7 (Print)
ISBN 978-951-41-1047-4 (PDF)

Received 25 January 2010

2000 Mathematics Subject Classification:
Primary 39A10; Secondary 30D35, 39A12.

YLIOPISTOPAINO
HELSINKI 2010

ACKNOWLEDGEMENTS

First, I wish to thank my advisor Dr Risto Korhonen who suggested this topic to me and patiently helped me with his advice during the preparation of this thesis. I also want to express my gratitude to Dr Rod Halburd for some of his comments and especially for giving me valuable insight into many parts of the field of difference equations. He is partially to be credited for the general idea behind my research as well.

I am greatly indebted to professor emeritus Ilpo Laine, my second advisor, who taught me the rudiments of complex analysis, introduced me to Nevanlinna theory, offered me the opportunity to work on this field of research, and provided several helpful comments on the content of this thesis.

I want to express my gratitude to professors Raimo Hietarinta and Alfred Ramani for their critical comments and valuable suggestions; both pointed out mistakes in the original manuscript the fixing of which resulted in much more insight on the subject matter. I sincerely thank professor Kazuya Tohge for carefully reading the manuscript and providing several improvements to many parts of the thesis. I would also like to thank Casey Jedlowski and Chris Rogers for reading through the manuscript and pointing out mistakes in the language.

Both the University of Joensuu and University College London have provided me with excellent working conditions, for which I am very grateful. I thankfully acknowledge the financial support from of the Academy of Finland (projects 118314, 217085, and 134792). My visit to London, during which a major part of the research was carried out, was supported financially by the Väisälä Fund of the Finnish Academy of Science and Letters, the Magnus Ehrnrooth Foundation, the Emil Aaltonen Foundation, and the North Karelia Regional Fund of the Finnish Cultural Foundation.

Finally, I collectively thank my family members, friends, and the staff of the Mathematics department of the late University of Joensuu for all support.

Joensuu, January 2010

Onni Ronkainen

Contents

1. Introduction

In the past two decades, the interest in nonlinear analytic difference equations has increased, especially in response to the programme of finding some kind of an analogue of the Painlevé property of differential equations (see for instance [12]) for difference equations. Despite – and in part, due to – several suggestions, it has become rather clear that such an analogue will not have the clean, short formulation of its model in the field of differential equations.

A great number of difference versions of the well-known six Painlevé equations have been identified, and they have been found to share various integrability properties. One large open problem is to construct a more systematic framework that would enable easier recognition and classification of such equations.

There is no clear agreement in the literature on the notion of 'integrability', but instead a number of different approaches leading to slightly different classifications of equations. Some classical properties are, however, quite generally accepted to indicate integrability, such as the existence of related linear problems, Bäcklund transformations, special solutions, and relations to lattice soliton equations [7, 8, 18].

The working hypothesis in this thesis is that the existence of sufficiently many meromorphic solutions of sufficiently slow growth for a given difference equation is an indicator of that equation being integrable. We will specify the exact meaning of "sufficiently slow" later. This is a well-defined complex analytic property, much like the original Painlevé property of differential equations.

A paper by Ablowitz, Halburd, and Herbst [1] can be considered a landmark in the application of value distribution theory in the study of difference equations. They observed that all of the relevant discrete difference equations have obvious analytic versions, and hence can be studied using the methods of complex analysis, and in particular those of Nevanlinna's theory of value distribution. More precisely, we can, instead of a sequence y_n where $n \in \mathbb{N}$, consider a meromorphic function $y(z)$, understanding that $y(z+1)$ corresponds to y_{n+1}. There are few results actually ensuring the existence of meromorphic solutions for a given nonlinear difference equation of order two or higher, but nonetheless such solutions generally seem to exist, which is in stark contrast to the case of differential equations.

The approach in [1] has been developed further and successfully applied to identify certain integrable difference equations [16]. The idea is to consider a certain family of equations, assume that there exists a solution that has sufficiently slow growth, and use Nevanlinna theory with some additional reasoning to reduce the family into a list of special equations that can have such solutions. In practice, some additional assumptions, which we will describe later, must be made.

In [16], the family of equations containing what are traditionally called the difference Painlevé I and II equations was considered. Our aim is to study in a similar fashion two other families: the ones of the difference Painlevé III and V. Our results are given in Chapter 3, where we also give an exact statement of the result of [16] for comparison.

A rather strong formal similarity between Nevanlinna theory and Diophantine approximation was observed independently by Osgood [34, 35] and Vojta [48], and through this analogy, another property, Diophantine integrability, that possibly indicates integrability in a purely discrete setting, has been found by Halburd [13]. This and other methods related to the use of value distribution theory and similar considerations in the study of integrability will be discussed in the final chapter.

The aim of the present discourse is not only to give the aforementioned results and thus strenghten and expand on the hypothesis first introduced in [1], but also to demonstrate the important role that value distribution theory can play in the study of integrable discrete systems. This role, although widely recognised, seems often to be greatly underestimated.

This thesis consists of essentially four parts. In Chapter 3 we shortly review the development of the theory of difference Painlevé equations and in particular the application of Nevanlinna theory to the study of these equations. We state our main results and several lemmas that will be used in proving them. We also discuss some of the difficulties arising in these proofs, and so it is strongly advised that the reader is familiar with Chapter 3 before proceeding to the proofs. Chapter 3 also introduces much notation that will be used throughout the rest of the thesis.

In Chapters 4 and 5 we prove our main results, treating the families of difference Painlevé III and V, respectively. Finally, in Chapter 6, we review our results, give some examples, and consider some of the open questions.

A very short introduction to Nevanlinna theory, for those readers not familiar, is given in the following chapter.

2. A Brief Outline of Nevanlinna Theory

Nevanlinna theory can be used to study the density of points in the complex plane at which a meromorphic function takes a prescribed value. It also provides a natural way to measure the growth of the function. Formulations of Nevanlinna theory exist in specialised as well as more general settings, but we only present here the traditional theory.

In this short introduction no proofs will be provided, and examples will be kept to a minimum. For a thorough representation of the theory, see for instance [4] or [21].

Given a function f meromorphic in the whole complex plane $\mathbb{C}$, we define the following three real functions: the *proximity function*

$$m(r,f) := \frac{1}{2\pi}\int_0^{2\pi} \log^+ |f(re^{i\theta})| d\theta,$$

where $\log^+ x := \max\{0, \log x\}$; the *(integrated) counting function*

$$N(r,f) := \int_0^r \frac{n(t,f) - n(0,f)}{t} dt + n(0,f)\log r,$$

where $n(r,f)$, the unintegrated counting function simply counts the poles (counting multiplicities) of f in a disc of radius r, centred at the origin; and finally the *characteristic function*

$$T(r,f) := m(r,f) + N(r,f).$$

The proximity function describes in some sense how close on average the values of f are to infinity on the circle $\{z \in \mathbb{C} \mid |z| = r\}$. The characteristic function provides a good representation of the complexity of f. The *order* of a meromorphic function f is defined as

$$\rho(f) := \limsup_{r\to\infty} \frac{\log T(r,f)}{\log r},$$

and this quantity is the same as the classical order (defined in terms of the maximum modulus) in the case that f is entire. We will, however, be dealing with functions that potentially have infinite order, and for them the notion of hyper-order (or first iterated order) is needed. The *hyper-order of* f is defined as

$$\rho_2(f) := \limsup_{r\to\infty} \frac{\log^+ \log^+ T(r,f)}{\log r}.$$

We will be interested in functions of hyper-order strictly less than one, as for them the Nevanlinna functions satisfy some useful inequalities (to be introduced later). The simplest example of a function that has hyper-order exactly one is $\exp(\exp(z))$.

As the definitions of the order and hyper-order suggest, we are often interested in the asymptotic behaviour of the Nevanlinna functions of a given meromorphic function. Many of the results of Nevanlinna theory hold only for most values of r. The set of all those values r where some property does not hold, is called an exceptional set. We will mostly deal with exceptional sets E of finite logarithmic measure, i.e. such that $\int_E \frac{1}{t} dt < \infty$.

Let g be another meromorphic function. In what follows, we will always assume functions to be meromorphic in $\mathbb{C}$. If $T(r,g) = o(T(r,f))$ as $r \to \infty$ outside an exceptional set of finite logarithmic measure, we say that g *is small with respect to* f. Any such small error term is denoted by $S(r,f)$, so that in this case we have $T(r,g) = S(r,f)$. Intuitively, this simply means that the characteristic of g is smaller than that of f for most values r. The family of all meromorphic functions that are small with respect to f is denoted by $\mathcal{S}(f)$.

We collect the most important basic properties of the characteristic function in the following:

(1) $T(r,f) = T(r, \frac{1}{f-a}) + O(1)$ for any $a \in \mathbb{C}$ (the first main theorem).
(2) $T(r,f) = O(\log r)$ if and only if f is rational.
(3) $T(r,f)$ is a convex increasing function of $\log r$.
(4) $T(r, f+g) \leq T(r,f) + T(r,g) + O(1)$ for any two functions f and g, as well as
(5) $T(r, fg) \leq T(r,f) + T(r,g)$.

The error terms in 1 and 4 can be expressed much more accurately (see e.g. [4]), but these rough bounds suffice for our purposes. The function $N(r,f)$ is also increasing, but the same cannot be guaranteed for $m(r,f)$. Inequalities 4 and 5 hold for $m(r,f)$ and $N(r,f)$ separately as well, and they can obviously be extended for a sum or a product of finitely many functions.

More properties of the characteristic function, especially those useful in the study of difference equations, will be introduced in the next chapter.

We have completely dismissed some of the most important results of Nevanlinna theory, like the logarithmic derivative lemma and the second main theorem, simply because they are not needed in the course of this thesis.

3. Difference Equations and Meromorphic Solutions

3.1. **Equations with constant coefficients.** We will see that we get much "nicer" results when we restrict the coefficients of the difference equations to be rational, in which case they are forced to be constants in almost all cases. Another extra restriction we will apply is to require certain coefficients to be periodic with period 1. In the completely discrete setting, when considering sequences instead of functions, a periodic coefficient with period 1 is just a constant, but in our analytic setting such a coefficient can quite well be a transcendental function.

Equations with constant coefficients have been studied widely, and they have many applications in several areas of applied mathematics, e.g. in the study of quantum gravity [6]. The equations that are now traditionally called the discrete Painlevé equations with constant coefficients are special cases of the QRT (Quispel-Robert-Thompson) difference equations. The QRT family was also a starting point in the discovery of the discrete Painlevé equations in [38]. The symmetric QRT family is

$$w(z+1) = \frac{f_1(w(z)) - w(z-1)f_2(w(z))}{f_2(w(z)) - w(z-1)f_3(w(z))}, \tag{3.1}$$

where the f_j, defined by

$$\begin{pmatrix} f_1(w) \\ f_2(w) \\ f_3(w) \end{pmatrix} = \begin{pmatrix} \alpha_0 & \beta_0 & \gamma_0 \\ \beta_0 & \delta_0 & \varepsilon_0 \\ \gamma_0 & \varepsilon_0 & \zeta_0 \end{pmatrix} \begin{pmatrix} w^2 \\ w \\ 1 \end{pmatrix} \times \begin{pmatrix} \alpha_1 & \beta_1 & \gamma_1 \\ \beta_1 & \delta_1 & \varepsilon_1 \\ \gamma_1 & \varepsilon_1 & \zeta_1 \end{pmatrix} \begin{pmatrix} w^2 \\ w \\ 1 \end{pmatrix}$$

are, in general, quartic polynomials. Equation (3.1) possesses generically a two-parameter family of finite-order meromorphic solutions, expressed in terms of elliptic functions (see for instance [18]).

3.2. **Earlier results involving Nevanlinna theory.** First order analytic difference equations have been studied by several authors. We mention in particular Yanagihara, who proved the following two results [50]. First, for any nonconstant rational function $R(w)$, the difference equation

$$w(z+1) = R(w)$$

has a nontrivial meromorphic solution. Second, if the equation

$$w(z+1) = R(z, w),$$

where R is rational in both z and w admits a transcendental meromorphic solution of finite order, then $\deg_w R = 1$; in other words the equation is a difference Riccati equation. Several other results exist in the first order case, which can be said to be well studied by now. Our focus will be on second order difference equations. Higher order equations have also been studied by Yanagihara, who introduced Nevanlinna theory for half-strip domains for this purpose (see [51] and references therein).

Definition 3.1. A meromorphic solution w of a difference equation is called *admissible* if all the coefficients of the equation are in $\mathcal{S}(w)$.

Remark. In particular, if the coefficients are rational, an admissible solution must be transcendental, and if the admissible solution itself is rational, then the coefficients must be constants.

The term "admissible" comes from the theory of complex differential equations, and should not be interpreted to mean that inadmissible solutions would somehow be unacceptable. What is meant is that the solution is admissible for the application of Nevanlinna theory.

It has been proposed that the existence of sufficiently many meromorphic solutions of finite order would be a strong indicator of integrability of an equation (see [1, 15, 16, 18]). In this claim, originally given by Ablowitz, Halburd, and Herbst in [1], no distinction is made between admissible and inadmissible solutions. In the actual theorems that have been proved, the additional requirement of admissibility is introduced, but solely due to the fact that existing tools in Nevanlinna theory are insufficient to handle the situation where solutions grow roughly at the same rate as the coefficients.

The existing results applying Nevanlinna theory to difference equations, as well as new results given in this thesis, do not consider the case where a difference equation has a large number of inadmissible solutions and no admissible ones of finite order. This should not be intepreted to mean that inadmissible solutions are not important. Some arguments (but no rigorous proofs) concerning inadmissible solutions were given in [11]. In the case of constant coefficients, it is possible to show that certain nonintegrable equations have only a finite number of constant (inadmissible) solutions, while all other solutions are of infinite order. Then, allowing nonconstant coefficients will transform the formerly constant solutions into inadmissible solutions which are still too few in number to call the equation integrable. In this light, the need to consider faster-growing (admissible) solutions is related to the too small number of very slow growing ones, not to any intrinsic "unacceptability" of the latter. This kind of argumentation also suggests that one might try to prove that there are, also in the general case, only a relatively small number of inadmissible solutions. This seems, however, to be a difficult task.

More recent results, including those in this thesis, would suggest that it might be reasonable to replace "finite order" with "hyper-order less than one". The essential point, that the solutions must not grow too fast, of course remains valid. Given a solution $w(z)$ for an autonomous complex difference equation, it is in a very general sense possible to construct a solution of arbitrarily high order by replacing the argument z by a suitable periodic entire function.

Equations of the form

$$w(z+1) \star w(z-1) = R(z, w),$$

where R is rational in both of its arguments and $\star$ stands for either the addition or the multiplication, were studied in [1], where it was shown that the existence of a nonrational meromorphic solution of finite order implies $\deg_w R \leq 2$. This class of equations contains many equations considered to be integrable, including the equations often called difference Painlevé I–III.

Notation. For convenience, we usually suppress the z-dependence by writing $f(z) = f$, $f(z+1) = \overline{f}$ and $f(z-1) = \underline{f}$ for any function f. For higher shifts we use $\overline{\overline{f}} = f(z+2)$, $\overline{f}^{[3]} = f(z+3)$, $\underline{f}_{[3]} = f(z-3)$, etc.

For brevity, a periodic function with period k ($k \in \mathbb{C}$, $k \neq 0$) will be called *a period-k function*, or just *period-k*.

The following theorem was originally proven in [16]. We present it here in a slightly generalised form, with the requirement "finite order" replaced by "hyper-order less than one":

Theorem 3.2 ([19]). *Assume that the equation*

$$\overline{w} + \underline{w} = R(z, w), \tag{3.2}$$

where $R(z,w)$ is rational in w and meromorphic in z, has a meromorphic solution w which is admissible in the sense of Definition 3.1 and has hyper-order less than *one. Then either w satisfies a difference Riccati equation*

$$\overline{w} = \frac{\overline{\alpha} w + \beta}{w + \alpha},$$

where $\alpha, \beta \in \mathcal{S}(w)$, or equation (3.2) can be transformed by a linear change in w to one of the following equations:

$$\overline{w} + w + \underline{w} = \frac{p_1 z + p_2}{w} + q_1, \tag{3.3a}$$

$$\overline{w} - w + \underline{w} = \frac{p_1 z + p_2}{w} + (-1)^z q_1, \tag{3.3b}$$

$$\overline{w} + \underline{w} = \frac{p_1 z + q_1}{w} + \frac{p_2}{w^2}, \tag{3.3c}$$

$$\overline{w} + \underline{w} = \frac{p_1 z + p_3}{w} + p_2, \tag{3.3d}$$

$$\overline{w} + \underline{w} = \frac{(p_1 z + q_1)w + p_2}{(-1)^{-z} - w^2}, \tag{3.3e}$$

$$\overline{w} + \underline{w} = \frac{(p_1 z + q_1)w + p_2}{1 - w^2}, \tag{3.3f}$$

$$\overline{w} w + w \underline{w} = \alpha, \tag{3.3g}$$

$$\overline{w} + \underline{w} = \alpha w + \beta, \tag{3.3h}$$

where $p_k, q_k \in \mathcal{S}(w)$ are arbitrary period-k functions.

All of the equations (3.3a)–(3.3h) are widely considered to be integrable, and the list includes all equations of Painlevé type in the class (3.2). See [16] for further details and references. Equation (3.3a) is historically known as the difference Painlevé I equation; equation (3.3f) as difference Painlevé II. A difference Riccati equation is explicitly solvable in the case of constant coefficients and can be transformed into a second order linear difference equation in the general case.

The work on the family $w(z+1)w(z-1) = R(z, w)$, which includes the so-called difference Painlevé III, was initiated in [18], where a certain subcase of this family of equations was considered with an additional assumption that the order of the poles of w is bounded.

Nevanlinna theory can be applied in a similar way to study q-difference equations, i.e. equations where the shift $f(z + c)$ is replaced with $f(qz)$, $q \in \mathbb{C} \setminus \{0\}$. See [2] for more details. Methods closely related to the Nevanlinna approach are discussed in the final chapter.

3.3. **The main results.** We will be studying equations of the form

$$L(w) = R(z, w), \tag{3.4}$$

where R is rational in w and meromorphic in z, and $L(w)$ is one of the following products:

$$\overline{w}\underline{w} \qquad \text{(the family of difference Painlevé III)},$$
$$(w\overline{w}-1)(w\underline{w}-1) \qquad \text{(the family of difference Painlevé V)}.$$

We will also denote

$$R(z,w)=\frac{P(z,w)}{Q(z,w)},$$

where P and Q are polynomials in w with degrees p and q, respectively. It is always assumed that P and Q have no common factors. Our results for these equation families, which will be proven in later chapters, are introduced next.

Definition 3.3. An *algebroid function* is an n-valued function f defined by an irreducible relation

$$A_nf^n+A_{n-1}f^{n-1}+\ldots+A_1f+A_0=0,$$

where A_j are entire functions.

Theorem 3.4. *Assume that the equation*

$$\overline{w}\underline{w}=R(z,w) \tag{3.5}$$

has a meromorphic solution w which is admissible in the sense of Definition 3.1 and has hyper-order less than one. Then either w satisfies a difference Riccati equation

$$\overline{w}=\frac{\alpha w+\beta}{w+\gamma}, \tag{3.6}$$

where α,β and γ are algebroid functions small with respect to w, or equation (3.5) can be transformed by $w\to\alpha w$ or $w\to\alpha/w$, where α is a small algebroid function, to one of the following equations:

$$\overline{w}\underline{w} = \frac{\eta w^2-\lambda w+\mu}{(w-1)(w-\nu)}, \tag{3.7a}$$

$$\overline{w}\underline{w} = \frac{\eta w^2-\lambda w}{w-1}, \tag{3.7b}$$

$$\overline{w}\underline{w} = \frac{\eta(w-\lambda)}{w-1}, \tag{3.7c}$$

$$\overline{w}\underline{w} = hw^m. \tag{3.7d}$$

In (3.7a), the coefficients satisfy $\kappa^2\overline{\mu}\underline{\mu}=\mu^2$, $\overline{\lambda}\mu=\kappa\underline{\lambda}\overline{\mu}$, $\kappa\overline{\overline{\lambda}}\underline{\lambda}=\underline{\kappa}\lambda\overline{\lambda}$, and one of the following:

$$(1)\ \eta\equiv 1,\ \overline{\nu}\underline{\nu}=1,\ \kappa=\nu; \qquad (2)\ \overline{\eta}=\underline{\eta}=\nu,\ \kappa\equiv 1.$$

In (3.7b), $\eta\overline{\eta}=1$ and $\overline{\overline{\lambda}}\underline{\lambda}=\lambda\overline{\lambda}$. In (3.7c), the coefficients satisfy one of the following:

(1) $\eta\equiv 1$ *and either* $\lambda=\overline{\lambda}\underline{\lambda}$ *or* $\overline{\lambda}^{[3]}\underline{\lambda}_{[3]}=\overline{\overline{\lambda}}\underline{\underline{\lambda}}$;

(2) $\overline{\lambda}\underline{\lambda} = \overline{\overline{\lambda}}\underline{\underline{\lambda}}$, $\overline{\eta}\overline{\lambda} = \overline{\overline{\lambda}}\underline{\eta}$, $\eta\underline{\eta} = \overline{\overline{\eta}}\underline{\eta}_{[3]}$;

(3) $\overline{\overline{\eta}}\underline{\underline{\eta}} = \eta\underline{\eta}$, $\lambda = \underline{\eta}$;

(4) $\overline{\lambda}^{[3]}\underline{\lambda}_{[3]} = \overline{\overline{\lambda}}\underline{\underline{\lambda}}\lambda$, $\eta\lambda = \overline{\overline{\eta}}\underline{\underline{\eta}}$.

In (3.7d), $h \in \mathcal{S}(w)$ and $m \in \mathbb{Z}$, $|m| \leq 2$.

Remark. Equivalently vanishing coefficients are allowed in all equations (3.7a)–(3.7c), as long as the required relations are satisfied. In principle it is also possible that the numerator of the right hand side of equation (3.7a) has $(w - \nu)$ as a factor. We have attempted to give for the coefficients such relations that make the above presentation as brief as possible. Observe that when $\mu \not\equiv 0$, the equation $\kappa\overline{\overline{\lambda}}\underline{\lambda} = \underline{\kappa}\lambda\overline{\lambda}$ given for λ in (3.7a) is redundant since it follows from the other two equations. A similar redundancy appears in option (2) for (3.7c).

In contrast to Theorem 3.2, we are unable to give explicit coefficients, and have instead only listed the conditions that the coefficients must satisfy. Most of the "small equations" appearing in the theorem have a large number of solutions, and even when a general solution could perhaps be obtained, it would be very difficult to express briefly; see Section 3.3.1 below. More accessible formulations of the equations (3.7a)–(3.7c) can be obtained if we restrict the coefficients; see Section 3.3.2. The connections of the equations in Theorem 3.4 to known integrable equations in the family (3.5) are discussed there, as well.

In our considerations the algebroid functions generally have square root type branch points. These algebroid functions arise from the factorisation of the polynomials P and Q as will be seen in the proof in Chapter 4. Section 3.4.2 below explains how we can deal with them in Nevanlinna theory.

Our treatment on the family of difference Painlevé V is much more complicated result-wise. There are so many different possible variations that it seems pointless to formulate a theorem in the same manner as we have done with the family of difference Painlevé III in Theorem 3.4. Instead, we state the results here only roughly, and refer the reader to the actual treatment in Chapter 5 for further details.

Let w be an admissible meromorphic solution of the equation

$$(w\overline{w} - 1)(w\underline{w} - 1) = \frac{P(z,w)}{Q(z,w)}. \tag{3.8}$$

If $\rho_2(w) < 1$, then either w satisfies a difference Riccati equation (3.6), or equation (3.8) simplifies to an equation where, on the right hand side, $p \leq 4$ and $q \leq 2$, and if $q = 2$, then $p = 4$, and if $q = 1$, then $p \geq 2$. We also find several restrictions on the coefficients in the polynomials P and Q. For precise restrictions on all of the coefficients, see the proof in Chapter 5 (in particular, pages 47–51).

Our results are greatly simplified if we assume that the roots of P are period-1, which corresponds to the case of them being constant in the completely discrete setting. In the existing literature on the family of the difference Painlevé V equation, these coefficients indeed are always constant.

3.3.1. *On solving nonlinear equations.* The purpose of this section is to give some solutions to the various "small equations" that appear in Theorem 3.4. We are interested in meromorphic solutions with hyper-order less than one.

We start by considering

$$\overline{f}\underline{f} = f^2. \tag{3.9}$$

Now it immediately follows that $p := \overline{f}/f$ is period-1. Whittaker has shown in [49, Theorem 5] that the first-order linear difference equation

$$\overline{F} = pF, \tag{3.10}$$

where p is any meromorphic function with order $\rho(p) < \infty$, has a meromorphic solution F such that $\rho(F) \leq \rho(p) + 1$, i.e. also of finite order. Using this one can solve (3.9) as follows [22]: Fix one such solution and denote it by $F(z,p)$. Then $q := f/F(z,p)$ is period-1. Conversely, given two period-1 functions p and q, fix a solution $F(z,p)$ of (3.10), and then $f = qF(z,p)$ solves (3.9).

Hence the finite order solutions of (3.9) can be written in the form $f = qF(z,p)$, where q and p are period-1 functions and $F(z,p)$ satisfies (3.10). However, to obtain full generality, one would have to extend Whittaker's result to the case that $\rho_2(p) < 1$.

In what follows, let p_k and q_k be arbitrary period-k functions, $c \in \mathbb{C}$, and u_k an arbitrary kth root of unity. One concrete solution of (3.9) is $f = p_1 c^z u_2^{z^2}$, which corresponds to p in (3.10) being the exponential function, and can also be obtained by taking logarithms in (3.9) and formally solving the obtained linear equation (on these methods, see for instance [32]). A generalisation to this is provided with the hyperbolic G-function of Ruijsenaars [42], which solves (3.10) when p is trigonometric.

The following equations can be treated in a similar way:

$$\overline{\overline{f}}\underline{f} = f\overline{f}, \tag{3.11a}$$

$$\overline{f}^{[3]}\underline{f}_{[3]} = \overline{\overline{f}}\underline{\underline{f}}, \tag{3.11b}$$

$$\overline{f}\underline{f} = \overline{\overline{f}}\underline{\underline{f}}, \tag{3.11c}$$

$$f\underline{f} = \overline{\overline{f}}\underline{f}_{[3]}. \tag{3.11d}$$

A general solution of finite order to (3.11a) is given by $f = q_2 F(z, p_1)$, where $\overline{\overline{F}} = p_1 F$. This can be seen as above; we skip the details. A concrete example solution is $f = p_2 c^z u_4^{z^2}$. For equation (3.11b), the method using Whittaker's result gives $f = q_5 F(z, p_1)$, where $\overline{F}^{[5]} = p_1 F$, and one concrete family of solutions is $f = p_5 c^z u_{10}^{z^2}$. Equation (3.11c) has the general finite-order solution $f = q_3 F(z, p_1)$, where $\overline{F}^{[3]} = p_1 F$, and an example solution is $f = p_3 c^z u_6^{z^2}$. Finally, for equation (3.11d) we have in general $f = q_3 F(z, p_2)$, where $\overline{F}^{[3]} = p_2 F$, and for example $f = p_2 p_3 c^z u_{12}^{z^2}$.

The example solutions we have given are of the type that often appears in the literature on the discrete Painlevé equations. Usually the coefficients given in the discrete setting are of the form ac^n, where a, c are constants. We have replaced the constant a by a periodic function, and the exponent n by our independent variable z. The roots of unity in the coefficients are something that only arise from the complex analytic reasoning. For example $(-1)^{n^2}$, $n \in \mathbb{N}$, is a period-2 function, while c^{z^2}, $c \neq 0, 1$, is never periodic with any period.

Next we look at a couple of equations that we have been unable to solve generally:

$$\overline{\overline{f}}\,\underline{\underline{f}} = f\underline{f}, \tag{3.12a}$$

$$\overline{f}^{[3]}\underline{f}_{[3]} = \overline{\overline{f}}\,\underline{\underline{f}}\,f. \tag{3.12b}$$

Any period-1 function satisfies (3.12a), but this is hardly a general solution. It might be that the constants 1 and 0 are the only solutions with hyper-order less than one for (3.12b). If this would be true, option (4) for equation (3.7c) in Theorem 3.4 could be removed.

We also encounter on several occasions the following two equations:

$$f\overline{f} = 1, \tag{3.13}$$

$$\overline{f}\underline{f} = f. \tag{3.14}$$

All solutions of (3.13) are period-2, since (3.13) implies $\overline{f\overline{f}} = 1$ and thus $f = f\overline{f\overline{f}} = \overline{\overline{f}}$, while all solutions of (3.14) are period-6: (3.14) implies $\overline{f} = \overline{\overline{f}}f = \overline{\overline{f}f}\,\underline{f}$ and thus $\overline{\overline{f}}\underline{f} = 1$, or $\overline{f}^{[3]} f = 1$. As with (3.13), this implies $f = \overline{f}^{[6]}$.

Equation (3.13) is solved by $f = p_2/\overline{p}_2$, and this is at least a general solution of finite order. Namely, suppose that f is a finite-order meromorphic solution to (3.13). Then, by Whittaker's result, there exists a finite-order meromorphic function g such that $f = \overline{g}/g$, and substituting this into (3.13) implies that g is period-2.

Analogously, given a period-6 function p_6, the function $p_6\overline{p}_6/\left(\overline{p}_6^{[3]}\overline{p}_6^{[4]}\right)$ satisfies (3.14), but the general solution is unknown. Equation (3.14) also has the trivial solution $f \equiv 0$, but in the occasions when we encounter (3.14) we cannot accept zero as a solution. In particular, the only period-1 (or rational) solutions to (3.13) and (3.14) are the constants ± 1, and 0, 1, respectively.

We point out that choosing, for instance, $p_2 = \sin(\pi z)$ results in $p_2/\overline{p}_2 \equiv -1$. On the other hand, choosing p_2 to be for example an elliptic function gives genuinely transcendental solutions.

Theorem 3.4 also contains a rescaled version of (3.13), namely $\overline{f}\underline{f} = 1$, or $f\overline{\overline{f}} = 1$. Generically, the equation $f\overline{f}^{[k]} = 1$ is solved by $f = p_{2k}/\overline{p}_{2k}^{[k]}$.

Solving all the aforementioned equations for f is not enough, however. As we saw in the statement of Theorem 3.4, we often have another function, call it now g, that satisfies some equation together with f. Consider first

$$\overline{g}f = \underline{g}\overline{f}, \tag{3.15}$$

where f satisfies (3.9). We suppose that the functions are not identically zero, and thus g satisfies equation (3.11a). Substituting the general solutions (of finite order) of (3.9) and (3.11a) into (3.15), it can be seen that $f = q_1F(z, p_1)$, $g = q_2G(z, p_1)$ (with $\overline{F} = p_1F$ and $\overline{\overline{G}} = p_1G$), i.e. instead of four arbitrary periodic functions we have just three. If f is in the special family of solutions, $f = p_1c^{2z}u_2^{z^2}$, we can directly solve g from (3.15), and obtain $g = p_2c^zu_2^{z(z+1)/2}$.

Next we look at

$$\overline{g}\overline{f} = \overline{\overline{f}}\underline{g}, \tag{3.16}$$

where f satisfies (3.11c), which implies that g satisfies (3.11d). If $f = p_3c^{2z}u_6^{z^2}$, we can solve g from (3.16): $g = \frac{p_2}{p_3}c^zu_6^{z(z+3)/2}$.

While working with the family (3.8) in Chapter 5, we also find some difference equations with two or more unknowns that we are unable to solve generally. Some solutions can be found, and clearly at least constant solutions exist, so that these equations are indeed reasonable restrictions on the coefficients of R.

3.3.2. *Extra restrictions on the coefficients.* Suppose now that we restrict the coefficients in (3.5) and (3.8) to be of the form p_kc^z, where p_k is period-k and $c \in \mathbb{C}$. Such coefficients, often further simplified so that p_k is constant, are commonly given in the existing literature on discrete Painlevé equations, and so this restriction, which seems rather arbitrary from a theoretical point of view, is understandable from the point of view of applications.

In the following, we list a few previously known equations that are included in the results of Theorem 3.4:

$$\overline{w}\underline{w} = \frac{\overline{p}_2w^2 + q_2c^zw + p_1c^{2z}}{(w-1)(w-p_2)}, \tag{3.17a}$$

$$\overline{w}\underline{w} = \frac{w^2 + p_2e^{i\pi z/2}c^zw + p_1c^{2z}}{w^2-1}, \tag{3.17b}$$

$$\overline{w}\underline{w} = \frac{p_1c^z}{w(w-1)}, \tag{3.17c}$$

$$\overline{w}\underline{w} = \frac{w(w+p_2c^z)}{w^2-1}. \tag{3.17d}$$

Equation (3.17a), contained in option (2) for equation (3.7a), is often referred to as the difference Painlevé III equation. Its discrete version was first derived using the method of singularity confinement in [38]. Singularity confinement will be discussed later in Chapter 6. The difference Painlevé III equation is known to possess several properties suggesting integrability: Schlesinger transforms [24], a Lax pair [36], and special discrete Riccati solutions [25]. Some special solutions in terms of rational and discrete Bessel functions were presented in [9], and a Bäcklund transformation was given in [44].

Equation (3.17b), contained in option (1) for equation (3.7a) with $\nu \equiv -1$, was already obtained by Halburd and Korhonen in [18], using similar methods as in this thesis. Equation (3.17c), contained in option (2), is related to the equation

$$\overline{w}\underline{w} = \frac{dw+1}{w^2}, \tag{3.18}$$

where d satisfies $\overline{\overline{d}}\underline{\underline{d}} = d^2$. This equation was found in [39]. Changing $w \to -1/(dw)$ in (3.18) results in

$$\overline{w}\underline{w} = \frac{f}{w(w-1)},$$

where $f = \left(\overline{d}\underline{d}d^2\right)^{-1}$, and this f satisfies (3.9), just like the respective coefficient in (3.17c) must.

Equation (3.17d) is related to the equation

$$\overline{w}\underline{w} = \frac{w^2-t^2}{w-1} \tag{3.19}$$

by the transformation $w \to p_2c^z/w$. Equation (3.19) was linked to the discrete Painlevé equations in [27].

For the family (3.8) of difference Painlevé V, we force another rather arbitrary requirement that, in the factorised form of P, all but the leading coefficient must be period-1. With these restrictions, we find for instance the following equations:

$$L(w) = \frac{g_1f_1c^{2z}(w-p_1)(w-1/p_1)(w-q_1)(w-1/q_1)}{(w-g_1c^z)(w-f_1c^z)}, \tag{3.20a}$$

$$L(w) = \frac{q_1c^z(w-p_1)(w-1/p_1)(w^2-1)}{w^2-q_1c^z}. \tag{3.20b}$$

Here $L(w)$ denotes $(w\overline{w}-1)(w\underline{w}-1)$, and g_1, f_1 are arbitrary period-1 functions. Otherwise we have used the same notation as above.

Equation (3.20a) is traditionally called the difference Painlevé V equation. To get the form first introduced in [38], one can use the change $y = (w+1)/2$. A Bäcklund transformation as well as a special solution for the discrete Painlevé V equation were given in [44]. In [45], Schlesinger transformations and some particular solutions were found. See also [37]. The autonomous version of equation (3.20b) was discovered in [27].

We also wish to point out that should we restrict the coefficients to be of order less than one (analogously, in some sense, with the requirement that the solution has hyper-order less than one), they would almost invariably reduce to constants. In fact only the arbitrary coefficients, not restricted by any equation (for example h in equation (3.7d)), could be nonconstant.

3.4. **Nevanlinna theory and difference equations.** A key tool in practically all recent papers applying Nevanlinna theory to the study of difference equations is the difference analogue of the logarithmic derivative lemma. Slightly different formulations of this result were obtained independently by Halburd and Korhonen [14, 17], and by Chiang and Feng [5]. Originally valid for functions of finite order, the result was recently generalised to hold for meromorphic functions with hyper-order less than one:

Theorem 3.5 ([19]). *Let w be a nonconstant meromorphic function with $\rho_2(w) = \rho_2 < 1$, $c \in \mathbb{C}$ and $\delta \in (0, 1-\rho_2)$. Then*

$$m\left(r, \frac{w(z+c)}{w(z)}\right) = o\left(\frac{T(r,w)}{r^{\delta}}\right),$$

for all r outside a set of finite logarithmic measure.

All the various results obtained using the finite-order version, many of which will be needed in this thesis, generalise to the case of hyper-order less than one as well. The error terms will look slightly different, but their exact form is of little relevance to us. It is clear that Theorem 3.5 cannot hold in general for functions with hyper-order one or more, since for example $f(z) = \exp(2^z)$ satisfies $\overline{f}/f = f$.

In the rest of this section, we list several results that will be applied frequently in the next two chapters.

Theorem 3.6 (Valiron-Mohon'ko identity [46, 33]). *Let w be a meromorphic function and $R(z,w)$ a function which is rational in w and meromorphic in z. If all the coefficients of $R(z,w)$ are small compared to w, then*

$$T(r, R(z,w)) = (\deg_w R)T(r,w) + S(r,w).$$

The Valiron-Mohon'ko identity plays an important role in the study of differential equations, as well. The proof can be found in [29, Theorem 2.2.5].

Definition 3.7. A *difference polynomial* of order n is an expression of the form

$$U(z,w) = \sum_{\{J\}} \alpha_J(z) \left(\prod_{j \in J} w(z+c_j) \right),$$

where $\{J\}$ is a collection of subsets of $\{1, \dots, n\}$, $\alpha_J \in \mathcal{S}(w)$, $\alpha_J \not\equiv 0$, $n \in J$ for at least one J, and $c_j \in \mathbb{C}$. For each J, we call the number of elements in J the *degree* of that term. The *degree* of a difference polynomial, $\deg_w H$, is the maximum of the degrees of its terms.

Next, we introduce a generalisation of the difference version of the Clunie lemma (see [14]) by Laine and Yang [31, Theorem 2.3]. The proof in [31] uses only the finite-order version of the difference analogue of the lemma on the logarithmic derivative, but by using Theorem 3.5 instead we get the following formulation. Another result of the same type (but for a different class of equations) is given in [28].

Theorem 3.8. *Let w, $\rho_2(w) = \rho_2 < 1$, be a transcendental meromorphic solution of a difference equation*

$$H(z,w)A(z,w) = B(z,w),$$

where H, A, and B are difference polynomials, and suppose that H has only one term of maximal degree. If $\deg_w B \le \deg_w H$, then

$$m(r, A(z,w)) = o\left(\frac{T(r,w)}{r^{1-\rho_2-\epsilon}}\right) + S(r,w),$$

where $\epsilon > 0$, and the exceptional set related to $S(r,w)$ is of finite logarithmic measure.

The following result on the Nevanlinna characteristic is essential in the study of the family (3.8):

Theorem 3.9 ([11]). *For three meromorphic functions f, g and h the Nevanlinna characteristic satisfies*

$$T(r, fg + gh + hf) \le T(r,f) + T(r,g) + T(r,h) + O(1).$$

Notation. We adopt, for brevity, the following unconventional notations. The symbol $\gtrdot S(r,w)$ means *for more than $S(r,w)$ points*, and the symbol $\lessdot S(r,w)$ means *for at most $S(r,w)$ points*.

For example, if we say that a condition $C(z_j)$ holds $\lessdot S(r,w)$ z_j, it means that the integrated counting function counting the points z_j for which $C(z_j)$ holds is of growth $S(r,w)$.

Furthermore, if a meromorphic function f has a pole of order n at $z_0 \in \mathbb{C}$, that is,

$$f(z) = \beta(z - z_0)^{-n} + O\left((z - z_0)^{-n+1}\right), \quad \beta \ne 0,$$

for all z in an open disc of some positive radius centred at z_0, we denote this by $f(z_0) = \infty^n$. Analogously, an a-point of order n is denoted by $f(z_0) = a + 0^n$. The latter is a short notation for

$$f(z) = a + \beta(z - z_0)^n + O\left((z - z_0)^{n+1}\right), \quad \beta \ne 0,$$

for all z in an open disc of some positive radius centred at z_0.

3.4.1. *Detection of hyper-order at least one.* In this section we give the basic tools for telling when the solution w must have hyper-order at least one. First, we have a result from real analysis characterising the concept of "hyper-order less than one" in a very practical way.

Theorem 3.10 ([19]). *Let $T : [0,\infty) \to [0,\infty)$ be a nondecreasing continuous function, and $s \in (0,\infty)$. If*

$$\limsup_{r\to\infty} \frac{\log\log T(r)}{\log r} = \rho_2 < 1,$$

and $\delta \in (0, 1-\rho_2)$, then

$$T(r+s) = T(r) + o\left(\frac{T(r)}{r^\delta}\right),$$

where r goes to infinity outside a set of finite logarithmic measure.

Lemma 3.11. *Let w be an admissible meromorphic solution of $L(z,w) = R(z,w)$, where L is a difference polynomial in w, while R is rational in w and meromorphic in z. Suppose that $m(r,w) = S(r,w)$. If there exists $k > 0$ and $\alpha < \deg_w R$ such that*

$$N(r, L(z,w)) \leq \alpha N(r+k, w) + S(r,w),$$

then $\rho_2(w) \geq 1$.

Proof. First, observe that $m(r,w) = S(r,w)$ implies that also $m(r, L(z,w)) = S(r,w)$. Now, by Theorem 3.6,

$$\begin{aligned}(\deg_w R)T(r,w) &= T(r, L(z,w)) + S(r,w) = N(r, L(z,w)) + S(r,w) \\ &\leq \alpha N(r+k, w) + S(r,w) \leq \alpha T(r+k,w) + S(r,w),\end{aligned}$$

and so by Theorem 3.10, $\rho_2(w) \geq 1$. □

In practice Lemma 3.11 means, roughly speaking, that if we can group most poles of $L(w)$ in (3.4) with nearby poles of w so that the number of poles of $L(w)$ divided by the number of poles of w, both counting multiplicities, is less than $\deg_w R$, then w will have hyper-order at least one.

If R has the highest possible degree (the meaning of this will become apparent once we start proving Theorem 3.4), we get a stronger result. Let ℓ denote the degree of $L(w)$ as a difference polynomial in w, i.e. $\ell = 2$ for (3.5) and $\ell = 4$ for (3.8). Denote by $N'(r,w)$ the counting function of a certain subset of all the poles of w. What exactly this subset is will be defined separately whenever the next lemma is used. To say that $N'(r,w) > cT(r,w)$ for some $c > 0$ outside an exceptional set is the same as to say that there are more than $S(r,w)$ poles of this type.

Lemma 3.12. *Let ℓ and $N'(r,w)$ be as above. Assume that $m(r,w) = S(r,w)$ in (3.4), $\deg_w R = \ell$, $N'(r,w) > cT(r,w)$ for some $c > 0$ outside a set of finite logarithmic measure, and that $N'(r, L(w)) < \alpha N'(r+k, w)$ for some $\alpha < \deg_w R$ and $k \geq 1$. Then $\rho_2(w) \geq 1$.*

Proof. As in the previous lemma, we have $m(r, L(w)) = S(r,w)$. Again by Theorem 3.6, and the assumption that $\deg_w R = \ell$,

$$\begin{aligned}(\deg_w R)T(r,w) &= \big(N(r, L(w)) - N'(r, L(w))\big) + N'(r, L(w)) + S(r,w) \\ &\leq \ell\big(N(r+1, w) - N'(r+1, w)\big) + \alpha N'(r+k, w) + S(r,w) \\ &< \ell N(r+1, w) + (\alpha - \ell)cT(r + \max\{1,k\}, w) + S(r,w) \\ &\leq (\ell + (\alpha-\ell)c)T(r+k, w) + S(r,w),\end{aligned}$$

and so by Theorem 3.10, $\rho_2(w) \geq 1$. □

The difference between Lemmas 3.11 and 3.12 is that in the latter, all except at most $S(r,w)$ poles of w must be grouped with $\deg_w R$ nearby poles of $L(w)$ to avoid a contradiction, while in the former the contradiction is avoided if there are more than $S(r,w)$ poles that can be grouped this way. The difficulty when the degree of R is too low (less than ℓ) is that the estimate $N(r, L(w)) \leq \ell N(r+1, w)$ is too rough.

On the other hand, Lemma 3.11 is very generic and by no means limited to the two special equations we are studying here, while in the proof of Lemma 3.12 we used the special form of $L(w)$. The idea in Lemma 3.11 is similar to a lemma used in [16] to prove Theorem 3.2, but we have included the extra assumption $m(r, w) = S(r, w)$, knowing that we can satisfy this requirement in all the relevant cases. Various results of the same type, but without this extra assumption, are given in [30].

3.4.2. *On the coefficients.*

Lemma 3.13 ([16]). *Let w be a meromorphic function with more than $S(r, w)$ poles (or c-points, $c \in \mathbb{C}$) counting multiplicities, and let $a_1, a_2, \ldots, a_n \in \mathcal{S}(w)$. Assume moreover that none of the functions a_i is identically zero. Denote by z_j the poles and zeros of the functions a_i (where j is in some index set), and let*

$$m_j := \max_{1 \leq i \leq n} \{l_i \in \mathbb{N} \mid a_i(z_j) = \infty^{l_i} \text{ or } a_i(z_j) = 0^{l_i}\}$$

be the maximal order of zeros and poles of the functions a_i at z_j. Then for any $\epsilon > 0$ there are at most $S(r, w)$ points z_j such that $w(z_j) = \infty^{k_j}$ (or $w(z_j) = c + 0^{k_j}$), where $m_j \geq \epsilon k_j$.

Because Lemma 3.13 is rather essential, and because its proof is quite short, we repeat the proof here:

Proof. We prove the claim in the case when w has more than $S(r, w)$ poles; for c-points a similar proof works. Suppose that $w(z_j) = \infty^{k_j}$ with $m_j \geq \epsilon k_j$, $\gg S(r, w)$ z_j. Denote by $N_{z_j}(r, w)$ the counting function of those poles of w which are in the set $\{z_j\}$, and by $N_\Sigma(r, a_i)$ the counting function of the poles of all a_i. Then, by assumption

$$\limsup_{r \to \infty} \frac{N_\Sigma(r, a_i)}{T(r, w)} \geq \limsup_{r \to \infty} \frac{\epsilon N_{z_j}(r, w)}{T(r, w)} > 0,$$

where r stays outside a set with finite logarithmic measure. This implies that at least one of the functions a_i has more than $S(r, w)$ poles, contradicting the assumption that $a_i \in \mathcal{S}(w)$. □

We will only be using Lemma 3.13 in the case when w has more than $S(r, w)$ poles. The lemma is often needed in the following considerations, because in principle it could happen that the coefficients in (3.4) have poles or zeros always when w has a pole. However, as the lemma implies, at most points the order of the pole of w is much greater than those of the coefficients, and this is enough for our purposes.

As an example of a situation where the lemma would be needed (this example is given in [16], as well), consider the gamma function Γ, which has a simple pole at $-n+1$ for all $n \in \mathbb{N}$. The order of Γ is one. We can construct a meromorphic function G which has a pole of order n^2 at the points $-n^2$ [21]. Then $\rho(G) \geq 3$, and so $\Gamma \in \mathcal{S}(G)$. Observe that there are no points where G has a pole and Γ does not.

If the multiplicities of the poles of w have a uniform upper bound, or if the coefficients of (3.4) have only finitely many zeros and poles, then each ϵ in the reasonings of the following chapters can be replaced by zero.

While we only consider meromorphic solutions of equations with meromorphic coefficients, in the course of the reasoning we must also handle equations with coefficients that might have some branch points, i.e. finite-valued algebroid functions. The results of the classical Nevanlinna theory cannot be applied to such functions, but instead we can, when needed, rely on the Selberg-Ullrich theory, the algebroid version of Nevanlinna theory (see for instance [26]), which studies meromorphic functions on a finitely sheeted Riemann surface. Thus, whenever the coefficients are such that branch points exist, $T(r,\cdot)$ and $N(r,\cdot)$ will denote the corresponding characteristic and counting functions of a finite-sheeted algebroid function. All algebroid functions we need to consider are small functions with respect to w and so the change of the underlying theory actually only affects the error term $S(r,\cdot)$. It needs to be redefined in terms of the algebroid characteristic, but since it will remain small with respect to $T(r,w)$, we can still denote it by $S(r,w)$.

Actually, since the estimates we make on the small coefficients do not in fact require the strong results of Nevanlinna theory in most cases, the instances where we really need to rely on the algebroid theory are quite few. Often, either the branch points do not really cause any problems, or the problems could be avoided by using some auxiliary meromorphic function. We have, however, chosen to simply rely on algebroid theory, since this greatly simplifies the required reasoning.

4. The Family of Difference Painlevé III

The purpose of this section is to prove Theorem 3.4. The proof is rather involved with numerous subcases. We have attempted to organise the proof so that subcases that are treated using similar methods are together as far as is convenient. This hopefully makes following the proof a little bit easier, while unfortunately making it somewhat difficult to trace exactly where we obtain each possible equation mentioned in Theorem 3.4.

4.1. Preliminaries.

Applying first Theorems 3.6 and 3.10 to (3.5), we get

$$(\deg_w R)T(r,w) = T(r,\overline{w}\underline{w}) + S(r,w) \leq 2T(r,w) + S(r,w),$$

so that $\deg_w R \leq 2$. The zero case being just a subcase of (3.7d), we suppose from now on that $\deg_w R = 1, 2$. The next lemma gives a sufficient condition that the function w should have a large number of poles. Recall that p and q denote the degrees of the polynomials P and Q in (3.5), respectively.

Lemma 4.1. *Assume that the hypotheses of Theorem 3.4 hold. If $q \geq 1$, then $m(r,w) = S(r,w)$.*

Proof. By the above reasoning, $\max\{p,q\} \leq 2$. Whenever the assumption holds, we can write (3.5) as

$$w^q\underline{w}\overline{w} = \Phi(z,w),$$

where $q \in \{1,2\}$ and $\Phi(z,w)$ is a difference polynomial in w with degree at most $q+1$. If w is transcendental, we can use Theorem 3.8 with $H(z,w) = w^q \underline{w}$ to conclude that $m(r,\overline{w}) = S(r,w)$.

If w is rational, then by its admissibility the coefficients must be constants (and w itself cannot be constant). If w has $\deg(w)$ poles, then it follows from the definitions of the Nevanlinna functions that $m(r,w) = O(1)$, so we may suppose that $w(\infty) = \infty$. Thus $w(z) \sim cz^m$, $m \in \mathbb{N}$, as $|z| \to \infty$. Then $w(z+1)w(z-1) \sim c^2 z^{2m}$, and $R(z,w) \sim Az^{(p-q)m}$, $A \neq 0$, so that we have $2 = p - q \leq p - 1 \leq 1$, a contradiction. □

Lemma 4.2. *Let b_j be any of the roots of the polynomials P and Q. Then*

$$N\left(r, \frac{1}{w - b_j}\right) = T(r,w) + S(r,w).$$

Proof. Note that if the respective order p or q is equal to two, the roots b_j can be algebroid. Equation (3.5) can be written in the form $\Psi(z,w) = 0$, where Ψ is a difference polynomial in w. Because P and Q have no common factors the function b_j cannot be a solution of this equation. We will apply [14, Corollary 3.4] (the difference analogue of the Mohon'ko lemma), reformulating it to hold for functions of hyper-order less than one and with algebroid coefficients (the proof in [14] needs to be re-written to use Theorem 3.5, but this can be done with just few adjustments), and obtain that

$$m\left(r, \frac{1}{w - b_j}\right) = S(r,w).$$ □

We will next consider separately the cases $q = 0, 1, 2$.

4.2. **Equations with** $q = 2$. We consider the cases where (3.5) is of one of the following forms:

$$\overline{w}\underline{w} = \frac{c(w-h_1)(w-h_2)}{(w-a_1)(w-a_2)}, \tag{4.1a}$$

$$\overline{w}\underline{w} = \frac{c(w-h_1)}{(w-a_1)(w-a_2)}, \tag{4.1b}$$

$$\overline{w}\underline{w} = \frac{c}{(w-a_1)(w-a_2)}, \tag{4.1c}$$

where $c \in \mathcal{S}(w)$ and the small functions h_1, h_2, a_1, a_2 are assumed algebroid, because they can in principle, due to factorisation, have square root type branch points. We assume, for now, that these rational expressions are square free and that neither of the functions a_1, a_2 vanishes identically. The cases where we allow $a_1 = a_2$, $h_1 = h_2$, or identically vanishing roots in Q are treated later in this section.

We will follow to some extent the reasoning in [18], where equation (4.1a) with meromorphic coefficients was considered with an additional assumption concerning the order of the poles of w. All the cases (4.1a)–(4.1c) will be considered simultaneously as far as that is possible. This means that all following statements that do not refer to a specific case indeed can be applied equally to any of the cases (4.1a)–(4.1c).

By Lemma 4.2, $w-a_m$ and $w-h_m$, $m=1,2$, have a large number of zeros, loosely speaking. In addition, Lemma 4.1 implies that w has a large number of poles, and this implies that so does $\overline{w}\underline{w}$, since by Theorem 3.5

$$\begin{aligned} m(r,\overline{w}\underline{w}) &= m\left(r, w^2\frac{\overline{w}\underline{w}}{w^2}\right) \leq m(r,w^2) + m\left(r,\frac{\overline{w}}{w}\right) + m\left(r,\frac{\underline{w}}{w}\right) \\ &\leq 2m(r,w) + S(r,w) = S(r,w), \end{aligned}$$

and by Theorem 3.6 and (4.1a)–(4.1c), $T(r,\overline{w}\underline{w}) = 2T(r,w) + S(r,w) \neq S(r,w)$.

By Lemma 3.13, given $\epsilon > 0$, there are at most $S(r,w)$ points z_j where $Q(z_j,w) = 0^{k_j}$, but where $\overline{w}\underline{w}$ has a pole of order greater than $(1+\epsilon)k_j$ or less than $(1-\epsilon)k_j$ due to poles or zeros of $P(z_j,w)$. The combined effect of all such points can be included in the error term, and so we only consider the rest of the zeros of Q in what follows.

For a point z_j where $w(z_j) = a_m(z_j)$, define

$$L(z_j,w) = (\ldots, z_j-1, z_j, z_j+1, \ldots)$$

to be the longest possible list of points such that each $z_j + 2n \in L(z_j,w)$ is a zero of $w - a_m$, $m=1,2$, and each $z_j+2n+1 \in L(z_j,w)$ is a pole of w.

Suppose that w has more than $S(r,w)$ poles that are not contained in any sequence $L(z_j,w)$. Let $N^*(r,w)$ be the integrated counting function counting only such poles; by assumption we have $N^*(r,w) \geq CT(r,w)$ for some $C > 0$ in a set of infinite logarithmic measure. By equations (4.1a)–(4.1c), $N^*(r,\overline{w}\underline{w}) = S(r,w)$, and so we get

$$\begin{aligned} 2T(r,w) &= \big(N(r,\overline{w}\underline{w}) - N^*(r,\overline{w}\underline{w})\big) + N^*(r,\overline{w}\underline{w}) + S(r,w) \\ &\leq 2\big(N(r+1,w) - N^*(r+1,w)\big) + N^*(r+1,w) + S(r,w) \\ &\leq (2-C)T(r+1,w) + S(r,w), \end{aligned}$$

which implies that $\rho_2(w) \geq 1$ by Theorem 3.10. Therefore all except at most $S(r,w)$ poles of w are in some sequence $L(z_j,w)$.

We will call the total number of zeros of $w-a_m$ in $L(z_j,w)$ divided by the total number of poles of w (both counting multiplicities) the *a_m/pole ratio* of the sequence. By Lemma 3.12, this ratio can be less than some $\alpha < 2$ only for at most $S(r,w)$ sequences. (Otherwise we would consider a counting function $N'(r,w)$ counting only the poles in sequences with a ratio less than α, and the lemma would yield a contradiction.)

Consider a sequence $L(z_j,w)$ that contains only one zero of $w-a_m$. Then there are one or two poles in that sequence. With one pole we would have $w(z_j) = a_m(z_j) + 0^{k_j}$ and $w(z_j+1) = \infty^{m_j}$ or $w(z_j-1) = \infty^{m_j}$, where $(1-\epsilon)k_j < m_j$. If there are two poles, the situation is the same except that now we have $w(z_j+1)w(z_j-1) = \infty^{m_j}$. In any case, in such a sequence the a_m/pole ratio is at most $1/(1-\epsilon)$. Hence all except at most $S(r,w)$ sequences $L(z_j,w)$ contain at least two zeros of $w-a_m$. This means that there must be at least $T(r,w)+S(r,w)$ points z_j such that $w(z_j+1)=\infty$ and one of the following holds:

$$w(z_j)=a_2(z_j) \quad \text{and} \quad w(z_j+2)=a_1(z_j+2), \tag{4.2}$$

$$w(z_j)=a_1(z_j) \quad \text{and} \quad w(z_j+2)=a_2(z_j+2), \tag{4.3}$$

$$w(z_j)=a_1(z_j) \quad \text{and} \quad w(z_j+2)=a_1(z_j+2), \tag{4.4}$$

$$w(z_j)=a_2(z_j) \quad \text{and} \quad w(z_j+2)=a_2(z_j+2). \tag{4.5}$$

Since $N\big(r,1/(w-a_m)\big) = T(r,w)+S(r,w)$ holds for both choices of $m=1,2$, exactly one of the following is true:

(i) Both (4.4) and (4.5) hold $\gtrdot S(r,w)$.
(ii) Both (4.2) and (4.3) hold $\gtrdot S(r,w)$; (4.4) and (4.5) hold $\lessdot S(r,w)$.
(iii) Relation (4.3) holds $\gtrdot S(r,w)$; (4.2), (4.4), and (4.5) hold $\lessdot S(r,w)$.
(iv) Relation (4.2) holds $\gtrdot S(r,w)$; (4.3)–(4.5) hold $\lessdot S(r,w)$.

In what follows, we will derive some consequences separately for the conditions (i)–(iv).

4.2.1. *Cases (i) and (ii).* We first assume that (i) is true, and consider equation (4.1a). Now (4.4) holds $\gtrdot S(r,w)$, and starting from the assumption that $w(z_j)=a_1(z_j)$, we get by (4.1a) that

$$w(z_j)w(z_j+2)=a_1(z_j)a_1(z_j+2)=\left(\frac{c(w-h_1)(w-h_2)}{(w-a_1)(w-a_2)}\right)(z_j+1)=c(z_j+1)$$

$\gtrdot S(r,w)$. We find that $c(z_j)=a_1(z_j-1)a_1(z_j+1)$ at more than $S(r,w)$ points z_j, and so in fact $c=\underline{a}_1\overline{a}_1$, since the coefficients are small with respect to w. Similarly, by starting from $w(z_j)=a_2(z_j)$, we get that $c=\underline{a}_2\overline{a}_2$, which implies $a_2=\nu a_1$ for some $\nu\not\equiv 1$ that satisfies $\overline{\nu}\underline{\nu}=1$.

When we consider equation (4.1b), we find that

$$w(z_j)w(z_j+2)=a_1(z_j)a_1(z_j+2)=\left(\frac{c(w-h_1)}{(w-a_1)(w-a_2)}\right)(z_j+1).$$

Since $w(z_j+1)=\infty$, the right hand side is equal to 0, and so $a_1(z_j)=0$, $\gtrdot S(r,w)$. This implies that $a_1\equiv 0$, contradicting our assumption. The same result is obtained similarly for (4.1c).

Case (ii) works exactly in the same way as the above detailed considerations, and we get the results listed later in Table 1 on page 27.

4.2.2. *Cases (iii) and (iv).* Assume that (iii) holds. All except at most $S(r,w)$ zeros of $w-a_m$ are in sequences $L(z_j,w)$ containing at least two such zeros. By condition (iii), there has to be, in fact, exactly one zero of $w-a_1$ and one zero of $w-a_2$, since otherwise (4.2), (4.4), or (4.5) would hold $\gtrdot S(r,w)$. On the other hand, we already saw that w has at most $S(r,w)$ poles outside these sequences. Therefore all except at most $S(r,w)$ poles of w must be contained in sequences of the form

$$(\infty^{l_{j-}}, a_1+0^{k_{j-}}, \infty^{m_j}, a_2+0^{k_{j+}}, \infty^{l_{j+}}).$$

Here we understand that if $l_{j\pm} < 0$, the corresponding endpoint of the sequence is a zero of order $|l_{j\pm}|$, and if $l_{j\pm} = 0$, it is some nonzero finite value. As before, by Lemma 3.13, we have the restrictions

$$(1-\epsilon)k_{j\pm} < l_{j\pm} + m_j < (1+\epsilon)k_{j\pm}, \tag{4.6}$$

for both choices of the $\pm$ sign. Denote

$$U := (w - a_1)(\overline{w} - \overline{a}_2). \tag{4.7}$$

We will show that U is a small function with respect to w. Because $m(r,w) = S(r,w)$ by Lemma 4.1, also $m(r,U) = S(r,w)$. From the definition of U and the fact that all but at most $S(r,w)$ poles of w are in sequences of the above form, it follows that if U has more than $S(r,w)$ poles, then there are more than $S(r,w)$ sequences where $l_{j\pm} \neq 0$.

Consider first only those sequences (or those indices j) where $l_{j-} > 0$. We may assume that $l_{j-}/m_j \geq s > 0$ for all such sequences, because otherwise the l_{j-}:s are bounded while the m_j:s are not, in which case these poles will only have a small effect (at most $S(r,w)$) on $N(r,U)$. The a_m/pole ratio for the sequences in consideration is

$$\frac{k_{j-} + k_{j+}}{m_j + l_{j-} + \max\{0, l_{j+}\}} < \frac{2m_j + l_{j-} + l_{j+}}{(1-\epsilon)(m_j + l_{j-} + \max\{0, l_{j+}\})},$$

where we have used (4.6). Take d such that

$$\frac{1 + s/2}{1 + s} < d < 1.$$

Then $d \in (1/2, 1)$. We can choose ϵ small enough, so that the a_m/pole ratio is at most $2d$. From the estimate for the ratio, an obvious bound for a fixed j is

$$\epsilon_j < 1 - \frac{2m_j + l_{j-} + l_{j+}}{2d(m_j + l_{j-} + \max\{0, l_{j+}\})}.$$

We choose $\epsilon = \inf_j \epsilon_j$. To check that this infimum is not zero it suffices to consider

$$\begin{aligned} 2d(m_j + l_{j-} + \max\{0, l_{j+}\}) - 2m_j - l_{j-} - l_{j+} &\geq (2d-2)m_j + (2d-1)l_{j-} \\ &\geq m_j(2d - 2 + (2d-1)s) > 0, \end{aligned}$$

with our choice of d (note that $m_j \geq 1$ for all j). Because the infimum cannot be zero, our ϵ is well defined. Thus we conclude that if $l_{j-} > 0$, then in such sequences the a_m/pole ratio is at most some $2d < 2$, whenever ϵ is small enough. Now, if there were more than $S(r,w)$ sequences such that $l_{j-} > 0$, Lemma 3.11 would imply that $\rho_2(w) \geq 1$.

Next, consider in a similar fashion only those sequences where $l_{j-} < 0$. This time we choose $s > 0$ such that $|l_{j-}|/m_j \geq s$ for all such indices j. If such s does not exist, these zeros again have only a small effect on $N(r,U)$, and we are done. We also change our choice of d into one satisfying

$$\max\left\{\frac{1}{2}, \frac{2-s}{2}\right\} < d < 1,$$

and again $d \in (1/2, 1)$. Now the a_m/pole ratio is

$$\frac{k_{j-} + k_{j+}}{m_j + \max\{0, l_{j+}\}} < \frac{2m_j + l_{j-} + l_{j+}}{(1-\epsilon)(m_j + \max\{0, l_{j+}\})},$$

and we find a bound

$$\epsilon_j < 1 - \frac{2m_j + l_{j-} + l_{j+}}{2d(m_j + \max\{0, l_{j+}\})}.$$

To show that $\epsilon = \inf_j \epsilon_j$ is positive, we write

$$2d(m_j + \max\{0, l_{j+}\}) - 2m_j - l_{j-} - l_{j+} \geq (2d-2)m_j - l_{j-} \geq m_j(2d - 2 + s) > 0.$$

Similarly as above, we conclude that if there were more than $S(r, w)$ sequences such that $l_{j-} < 0$, then w would be of hyper-order at least one.

We can obviously repeat the above reasoning for the sequences where $l_{j+} \neq 0$. Therefore, we have shown that in the only noncontradictory cases $U \in \mathcal{S}(w)$, and so (4.7) becomes the Riccati difference equation (3.6).

The same reasoning works for case (iv) when we just change the roles of a_1 and a_2.

All the results obtained above are collected in Table 1. We found three kinds of results: restrictions on the coefficient functions (written down in the table); a contradiction; or that w satisfies a Riccati difference equation (3.6), which is denoted by (R) in the table.

TABLE 1. results obtained in the cases (i)–(iv)

Eq.	(i)	(ii)	(iii)	(iv)
(4.1a)	$c = \underline{a}_1\overline{a}_1 = \underline{a}_2\overline{a}_2$,	$c = \underline{a}_1\overline{a}_2 = \underline{a}_2\overline{a}_1$	(R)	(R)
(4.1b)	contradiction	contradiction	(R)	(R)
(4.1c)	contradiction	contradiction	(R)	(R)

By the results in Table 1, the forms (4.1b) and (4.1c) with $a_1 \neq a_2$ and both a_m not identically zero are now clear: they both lead to w satisfying a Riccati difference equation (3.6), or to a contradiction. For equation (4.1a) we still need some additional considerations.

To this end, we write $w = 1/y$, and assuming that h_1, h_2 are not identically zero, we can then write equation (4.1a) in the form

$$\overline{y}\underline{y} = \frac{\frac{a_1 a_2}{c h_1 h_2}(y - 1/a_1)(y - 1/a_2)}{(y - 1/h_1)(y - 1/h_2)}. \tag{4.8}$$

This is still of the form (4.1a), only with different coefficients. We can repeat the preceeding reasoning with the new coefficients, formulate an analogous set of conditions (i')–(iv'), and obtain the results in Table 2. The conditions (i')–(iv') read exactly as the conditions (i)–(iv) above, except that in equations (4.2)–(4.5) we must replace a_1, a_2 with $1/h_1, 1/h_2$.

TABLE 2. results obtained in the cases (i')–(iv')

Eq.	(i')	(ii')	(iii')	(iv')
(4.8)	$\frac{ch_1h_2}{a_1a_2}=\underline{h}_1\overline{h}_1=\underline{h}_2\overline{h}_2$	$\frac{ch_1h_2}{a_1a_2}=\underline{h}_1\overline{h}_2=\underline{h}_2\overline{h}_1$	(R)	(R)

In what follows, we apply the results of Tables 1 and 2 to equation (4.1a). We begin by making the transformation $w \to wa_1$, which gives

$$\overline{w}\underline{w}=\frac{\frac{c}{\overline{a}_1\underline{a}_1}\left(w^2-\frac{h_1+h_2}{a_1}w+\frac{h_1h_2}{a_1^2}\right)}{(w-1)\left(w-\frac{a_2}{a_1}\right)}. \tag{4.9}$$

Now we will go through all possible combinations of the conditions (i)–(iv) and (i')–(iv'). Observe that (iii), (iv), (iii'), and (iv') all lead to w satisfying a Riccati difference equation (if y satisfies a Riccati equation, then $w=1/y$ satisfies one, too, only with different coefficients), so we only need to consider the combinations of the rest of the cases.

Assume that (i) and (i') hold. Then, $a_2=\nu a_1$ with $\overline{\nu}\underline{\nu}=1$ and $h_1=\sigma h_2$ with $\overline{\sigma}\underline{\sigma}=1$. Applying the results in the tables, equation (4.9) becomes

$$\overline{w}\underline{w}=\frac{w^2-\frac{h_2(\sigma+1)}{a_1}w+\frac{\sigma h_2^2}{a_1^2}}{(w-1)(w-\nu)}.$$

On the other hand,

$$\overline{h}_2\underline{h}_2=\frac{ch_1h_2}{a_1a_2}=\frac{\sigma\overline{a}_1\underline{a}_1h_2^2}{\nu a_1^2},$$

and so the function $f=h_2/a_1$ satisfies $\nu\overline{f}\underline{f}=\sigma f^2$. Using this, one can verify that $\mu=\sigma f^2$ and $\lambda=f(\sigma+1)$ satisfy $\nu^2\overline{\mu}\underline{\mu}=\mu^2$ and $\nu\underline{\lambda}\overline{\mu}=\overline{\lambda}\mu$. This means that we have found equation (3.7a) in Theorem 3.4 with the option (1) and $\nu\not\equiv 1$.

Assume that (i) and (ii') hold. Combining the relations $c=\overline{a}_1\underline{a}_1=\overline{a}_2\underline{a}_2$ and $\frac{ch_1h_2}{a_1a_2}=\overline{h}_1\underline{h}_2=\overline{h}_2\underline{h}_1$ given in Tables 1 and 2 yields

$$\left(\frac{h_1h_2}{a_1a_2}\right)^2=\frac{\overline{h}_1\overline{h}_2}{\overline{a}_1\overline{a}_2}\cdot\frac{\underline{h}_1\underline{h}_2}{\underline{a}_1\underline{a}_2}, \tag{4.10}$$

Denote $\mu:=h_1h_2/a_1^2$. Then (4.10) can be written as $\nu^2\overline{\mu}\underline{\mu}=\mu^2$.

The following three equations hold $>S(r,w)$:

$$y(z_j+1)y(z_j-1)=\frac{a_1(z_j)a_2(z_j)}{c(z_j)h_1(z_j)h_2(z_j)}, \tag{4.11}$$

$$\left(y(z_j+1)-\frac{1}{h_1(z_j+1)}\right)\left(y(z_j+1)-\frac{1}{h_2(z_j+1)}\right)=0, \tag{4.12}$$

$$\left(y(z_j-1)-\frac{1}{h_1(z_j-1)}\right)\left(y(z_j-1)-\frac{1}{h_2(z_j-1)}\right)=0. \tag{4.13}$$

To see that the first one holds, recall that condition (ii') says that the following two relations hold $\gtrdot S(r,w)$:

$$y(z_j-1) = 1/h_2(z_j-1) \quad \text{and} \quad y(z_j+1) = 1/h_1(z_j+1),$$
$$y(z_j-1) = 1/h_1(z_j-1) \quad \text{and} \quad y(z_j+1) = 1/h_2(z_j+1).$$

Combining these with the result for (ii') in Table 2 gives (4.11). The other two equations, (4.12) and (4.13), are obtained directly from equation (4.8). From (4.11)–(4.13) and the result for (ii') in Table 2 we can deduce, after some manipulation, that $\gtrdot S(r,w)$

$$\text{(4.14)} \quad \Bigg((h_1(z_j+1)+h_2(z_j+1))\frac{c(z_j)h_1(z_j)h_2(z_j)}{a_1(z_j)a_2(z_j)} - (h_1(z_j-1)+h_2(z_j-1))h_1(z_j+1)h_2(z_j+1)\Bigg) y(z_j-1) = 0.$$

The coefficient of $y(z_j-1)$ must then vanish identically. Using the relations implied by (i), we get

$$\text{(4.15)} \qquad \frac{\overline{h}_1+\overline{h}_2}{\overline{a}_1}\cdot\frac{h_1h_2}{\nu a_1^2} = \frac{\underline{h}_1+\underline{h}_2}{\underline{a}_1}\cdot\frac{\overline{h}_1\overline{h}_2}{\overline{a}_1^2},$$

If we denote $\lambda := (h_1+h_2)/a_1^2$, this reads $\overline{\lambda}\mu = \nu\underline{\lambda}\overline{\mu}$, and we have again equation (3.7a), option (1), $\nu \not\equiv 1$.

Assume that (ii) and (i') hold. Then by Table 1, $c = \underline{a}_1\overline{a}_2 = \underline{a}_2\overline{a}_1$, which implies that $\nu := a_2/a_1$ is a period-2 function. The function $f = h_2/a_1$ satisfies now $\nu\overline{f}\underline{f} = \sigma\overline{\nu}f^2$, where σ is as above: $h_2 = \sigma h_1$ and $\overline{\sigma}\underline{\sigma} = 1$. It is straightforward to check that the functions $\lambda = \overline{\nu}f(\sigma+1)$ and $\mu = \overline{\nu}\sigma f^2$ satisfy the relations $\underline{\lambda}\overline{\mu} = \overline{\lambda}\mu$ and $\overline{\mu}\underline{\mu} = \mu^2$. This is option (2) for equation (3.7a) with the additional demand that $\nu \not\equiv 0,1$.

Assume that (ii) and (ii') hold. Combining the relations given in Tables 1 and 2 for (ii) and (ii') we again obtain equation (4.10): the function $g = h_1h_2/(a_1a_2)$ satisfies $\overline{g}\underline{g} = g^2$. The reasoning above which yields (4.14) is only dependent on the condition (ii'), and therefore it can be repeated here without any change. When we equate the coefficient of $y(z_j-1)$ in (4.14) to zero and apply the fact that $c = \overline{a}_2\underline{a}_1$, given by (ii) in Table 1, we get, instead of (4.15), the following:

$$\frac{\overline{h}_1+\overline{h}_2}{\overline{a}_1}\cdot\frac{h_1h_2}{a_1a_2} = \frac{\underline{h}_1+\underline{h}_2}{\underline{a}_1}\cdot\frac{\overline{h}_1\overline{h}_2}{\overline{a}_1\overline{a}_2}.$$

Denoting again $(h_1+h_2)/a_1$ by λ we have $\overline{\lambda}g = \underline{\lambda}\overline{g}$.

Since $a_2 = \nu a_1$ for a period-2 function ν, $\mu = \overline{\nu}h_1h_2/a_1^2$ is equal to $\overline{\nu}\nu g$. By the result for g, μ again satisfies $\overline{\mu}\underline{\mu} = \mu^2$. We also obtain the relation $\underline{\lambda}\overline{\mu} = \overline{\lambda}\mu$ by combining $\mu = \overline{\nu}\nu g$ and $\underline{\lambda}/\overline{\lambda} = g/\overline{g}$. Thus we have again obtained option (2) for equation (3.7a) with $\nu \not\equiv 0,1$.

We have now finished treating equations (4.1a)–(4.1c) with the assumptions that $a_1 \not\equiv a_2$, $a_m \not\equiv 0$, and in (4.1a), $h_1 \not\equiv h_2$ and $h_m \not\equiv 0$.

4.2.3. *Vanishing coefficients.* Next we consider the cases where one of the coefficients in $R(z,w)$ vanishes identically. We still suppose that $a_1 \not\equiv a_2$ and $h_1 \not\equiv h_2$.

Suppose that $a_1 \equiv 0$ (the reasoning is the same if $a_2 \equiv 0$). If $p = 0$, we write the equation as

$$\overline{w}\underline{w} = \frac{c}{w(w-a)}. \tag{4.16}$$

Starting from the assumptions that $w(z_j) = a(z_j) + 0^{k_j}$ and $w(z_j+1) = \infty^{m_j}$, we find that $w(z_j+2) = 0^{2m_j}$ and $w(z_j+3) = \infty^{m_j}$, where the order is given modulo a small error from the poles and zeros of the coefficients. The next iterate is

$$w(z_j+4) = \frac{a(z_j+2)^2 c(z_j+3) c(z_j+1)}{a(z_j) c(z_j+2)^2}.$$

If this is equal to $a(z_j+4)$ with a sufficiently large order, and this holds for $\gg S(r,w)$ points z_j, we conclude that $\overline{\overline{a}}\underline{\underline{a}}c^2 = a^2\overline{c}\underline{c}$. Otherwise, $w(z_j+5) = 0^{m_j}$ and $w(z_j+6) = \infty^{m_j}$, so that the a_m/pole ratio is less than one (observe that zeros of w are now zeros of $w - a_m$), leading to a contradiction. Note that this follows regardless of the iterates $w(z_j - n)$ in the negative direction, since the sum of the orders of the poles of w at z_j+1 and z_j-1 must be roughly equal to k_j (allowing also "poles of negative order", i.e. zeros).

Change $w \to aw$ in (4.16) and denote $\mu := c/(\overline{a}\underline{a}a^2)$. The relation obtained above can be written as $\mu^2 = \overline{\mu}\underline{\mu}$, and thus this case yields option (2) for equation (3.7a), with $\eta, \nu, \lambda \equiv 0$.

Let then $p = 1$, and write the equation as

$$\overline{w}\underline{w} = \frac{c(w-h)}{w(w-a)}. \tag{4.17}$$

Starting from the same assumptions as above, we find that $w(z_j+3) = 0^{m_j}$, and

$$w(z_j+4) = \frac{h(z_j+2) a(z_j) c(z_j+2)}{a(z_j+2) c(z_j+1)}.$$

If this equals $h(z_j+4)$ sufficiently often, we obtain $h\underline{\underline{a}}c = \overline{h}a\underline{c}$. If not, a contradiction is obtained as above. Changing $w \to 1/y$ in (4.17) gives an equation of the same form, but with different coefficients. Repeating the reasoning for this equation yields another condition, $\underline{a}\underline{\underline{h}}c = \overline{a}\underline{ch}$. Denote $\lambda := -c/(\overline{a}\underline{a}a)$ and $\mu := \lambda h/a$. Then we can deduce, using the obtained relations, that $\overline{\mu}\underline{\mu} = \mu^2$ and $\underline{\lambda}\overline{\mu} = \overline{\lambda}\mu$. This results in option (2) for equation (3.7a), with $\eta, \nu \equiv 0$.

Lastly, if $p = 2$, we change $w \to 1/y$, which gives an equation of the form (4.1b) with non-zero coefficients. This case has been treated above, and it lead to either w satisfying a Riccati difference equation or to a contradiction.

If $h_1 \equiv 0$ (or similarly $h_2 \equiv 0$) in (4.1a), we do the change $w \longrightarrow 1/y$. It gives the equation

$$\overline{y}\underline{y} = \frac{-\frac{a_1 a_2}{ch_2}(y - 1/a_1)(y - 1/a_2)}{y - 1/h_2}, \tag{4.18}$$

which is of the form where $q = 1$. The consideration of this case will be finished in Section 4.4.

The case that $h_1 \equiv 0$ in (4.1b) needs no separate consideration; a zero root in the numerator of $R(z, w)$ does not affect the reasoning leading to the conditions (i)–(iv), and from any of these it follows that $a_m \equiv 0$ for at least one value of m, contradicting the fact that P and Q have no common factors.

4.2.4. *The reduced cases.* We still have to see what happens if $a_1 = a_2$ or $h_1 = h_2$ or both in (4.1a)–(4.1c). First, we assume that a_1, a_2, h_1, h_2 are all not identically zero.

Assume that $a_1 = a_2$. This means that (4.4) holds for all the equations. (The reasoning needed to obtain (4.4) is only a slight modification of the original one.) Since we assume for now that $a_1, a_2 \not\equiv 0$, the only changing points in the above reasoning are in the consideration of equation (4.1a). Making the transformation $w \longrightarrow wa_1$ gives (4.9) with $a_2/a_1 = 1$.

Changing $w \longrightarrow 1/y$ gives an equation of the form (4.1a) with $h_1 = h_2$. This means that either of the following is true (compare to Table 2):

$$\text{(i'):}\ \frac{ch_1h_2}{a_1^2} = \overline{h}_1\underline{h}_1 = \overline{h}_2\underline{h}_2, \quad \text{(ii'):}\ \frac{ch_1h_2}{a_1^2} = \overline{h}_1\underline{h}_2 = \overline{h}_2\underline{h}_1.$$

Suppose that (i') is true, so that $\overline{f}\underline{f} = \sigma f^2$ for $f = h_2/a_1$ and $\overline{\sigma}\underline{\sigma} = 1$ (recall that σ is such that $h_1 = \sigma h_2$; cf. Subsection 4.2.1). Then $\lambda = f(\sigma + 1)$ and $\mu = \sigma f^2$ again satisfy $\overline{\mu}\underline{\mu} = \mu^2$ and $\underline{\lambda}\overline{\mu} = \overline{\lambda}\mu$. This is option (2) for equation (3.7a) with $\eta, \nu \equiv 1$. The same result is obtained in case (ii'), in which we have $p_2\overline{f}\underline{f} = \overline{p}_2 f^2$ for a period-2 function p_2 and $\lambda = f(p_2 + 1)$, $\mu = p_2 f^2$.

The case $h_1 = h_2$ needs no separate consideration, as the result can be transformed to the one obtained above with $w \rightarrow 1/y$.

The case that both $a_1 = a_2$ and $h_1 = h_2$ are true works in the same way, except for the simplification that $\sigma \equiv 1$.

Suppose then that $a_1, a_2 \equiv 0$. Changing w to $1/y$ in (4.1a) gives an equation of the form (4.1c), so that by Table 1 either y (and thus w as well) satisfies a Riccati difference equation, or we get a contradiction. The same change in (4.1b) also gives the form (4.1c), but with one root of Q identically zero; this case has been treated above. From equation (4.1c) we directly obtain (3.7d).

Finally, if $h_1, h_2 \equiv 0$, we transform $w \longrightarrow wa_1$ in (4.1a) and obtain in case (i)

$$\overline{w}\underline{w} = \frac{w^2}{(w-1)(w-\nu)}, \quad \overline{\nu}\underline{\nu} = 1, \tag{4.19}$$

a special case of option (1) for equation (3.7a), and in case (ii)

$$\overline{w}\underline{w} = \frac{\overline{\nu} w^2}{(w-1)(w-\nu)}, \qquad \overline{\nu} = \underline{\nu},$$

a special case of option (2). If $a_1 = a_2$, we get (4.19) with $\nu \equiv 1$.

The equations with $q = 2$ have now been handled, except for the case of (4.1a) with one root of P identically zero, which lead to (4.18). The results we found were of the form (3.7a), with option (1) or (2) in Theorem 3.4.

4.3. **Equations with $q = 0$.** We have two equations to consider, namely,

$$\overline{w}\underline{w} = c(w-h_1)(w-h_2), \quad \overline{w}\underline{w} = c(w-h_1). \tag{4.20}$$

If we allow the possibility $h_1 = h_2$, then these equations cover all the possible forms where $q = 0$. In the first equation h_1 and h_2 may be algebroid.

Assuming $h_1, h_2 \not\equiv 0$ and changing w into $1/y$, equations (4.20) are changed into

$$\overline{y}\underline{y} = \frac{\frac{1}{ch_1h_2} y^2}{(y-1/h_1)(y-1/h_2)}, \quad \overline{y}\underline{y} = \frac{-\frac{1}{ch_1} y}{y-1/h_1}. \tag{4.21}$$

The first one is of the form (4.1a) with $h_1, h_2 \equiv 0$; this was treated at the end of the previous section. The second equation in (4.21) has $q = 1$, so it will be treated in the next section.

If h_2 (or similarly h_1) vanishes, the first equation in (4.20) gives

$$\overline{y}\underline{y} = \frac{-\frac{1}{ch_1} y^2}{y-1/h_1}, \tag{4.22}$$

when we apply the change of w into $1/y$. This equation is of the form where $p = 2$, $q = 1$, and will be dealt with in the next section. If h_1 vanishes in the second equation in (4.20), or if both h_1 and h_2 vanish in the first one, we just get (3.7d).

4.4. **Equations with $q = 1$.** The powerful method of considering the a_m/pole ratio does not work in the case that $\deg_w R = 1$. Here we need a more careful inspection of the iterates of equation (3.5), and will use Lemma 3.11 to obtain contradictions. The considerations get rather lengthy, and we will not repeat needlessly any reasoning that has already been done above.

The equation we consider is

$$\overline{w}\underline{w} = \frac{P(z,w)}{w-a},$$

using our previous notation, but with a now definitely meromorphic. We still know that $w - a$ has a large number of zeros (Lemma 4.2), and that w has a large number of poles (Lemma 4.1). We will iterate the various forms of the equation starting from the assumptions that $w(z_j) = a(z_j) + 0^{k_j}$ and $w(z_j \pm 1) = \infty^{m_{j\pm}}$. Recall that, discarding at most $S(r, w)$ points, we have

$$(1-\epsilon)k_j < m_{j-} + m_{j+} < (1+\epsilon)k_j.$$

The iteration must be done separately for the cases when some of the coefficients vanish identically. Moreover, the results of the iteration are essentially different in the three different scenarios that $m_{j\pm} > 0$, $m_{j\pm} = 0$, and $m_{j\pm} < 0$. When considering the points where $m_{j\pm}$ is nonzero, we can assume that $|m_{j\pm}|/k_j$ is bounded from below by some positive constant, because otherwise the combined effect of all such points on the counting function can be included in the error term.

Case $p = 1$. The equation in consideration is now

$$\overline{w}\underline{w} = \frac{c(w-h)}{w-a}. \tag{4.23}$$

Suppose for now that neither a nor h is identically zero. We will next derive all the various forms of equation (3.7c) in Theorem 3.4.

Let $m_{j+} > 0$, $\gtrdot S(r, w)$. Then,

$$w(z_j + 2) = \frac{c(z_j + 1)}{a(z_j)}.$$

If this is not a zero of $w - a$ with a sufficiently high order, the next iterate $w(z_j + 3)$ will be zero, and

$$w(z_j + 4) = \frac{c(z_j + 3)h(z_j + 3)a(z_j)}{a(z_j + 3)c(z_j + 1)}.$$

If this is not a zero of $w - h$ with a sufficiently high order, the next iterate $w(z_j + 5)$ is infinite, and

$$w(z_j + 6) = \frac{a(z_j + 3)c(z_j + 5)c(z_j + 1)}{h(z_j + 3)a(z_j)c(z_j + 3)}.$$

The poles at $z_j + 1$ and $z_j + 5$ have the same order m_{j+} modulo a small error from the coefficients, so that we have so far k_j zeros of $w - a$ and roughly $2m_{j+}$ poles per sequence, unless one of the following is true:

$$\begin{aligned} &(\mathrm{A}): && c = \overline{a}\underline{a}, \\ &(\mathrm{B}): && \overline{c}\overline{h}\underline{\underline{a}} = \overline{\overline{h}}\overline{a}\underline{c}, \\ &(\mathrm{C}): && a\overline{\overline{c}}\underline{\underline{c}} = \overline{a}^{[3]}\underline{a}_{[3]}hc. \end{aligned}$$

The five iterates in the other direction can give at most (roughly) m_{j-} more zeros of $w - a$, but at the expense of yielding also (roughly) m_{j-} poles of w. In the case that $m_{j-} = 0$, we might, at worst, also have to look at the seventh iterates (if $z_j - 6$ is a zero of $w - a$), but this does not change our conclusion: unless one of (A)–(C) is true, the ratio of the zeros of $w - a$ and the poles of w in these sequences is strictly less than one, which yields a contradiction by Lemma 3.11.

We can consider $y = 1/w$ in a similar fashion, and find the conditions

$$\begin{aligned}
&\text{(A'):} && \frac{a}{ch} = \frac{1}{\overline{h}\underline{h}}, \\
&\text{(B'):} && \overline{\overline{a}}\underline{ch} = \underline{a}\overline{c}\underline{\underline{h}}, \\
&\text{(C'):} && \overline{\overline{a}}\underline{\underline{a}}\overline{h}^{[3]}\underline{h}_{[3]}c = \overline{\overline{c}}\overline{\overline{h}}\underline{\underline{ch}}.
\end{aligned}$$

One of (A)–(C), and one of (A')–(C') must be true. Here we assumed that $m_{j+} > 0$, but we need not consider separately the other, symmetrical possibility. Namely, conditions (A) and (A'), (C) and (C') are already symmetrical, and option (B) translated to the negative direction is just (B'), and vice versa.

Now we change $w \to wa$ in (4.23), obtaining

$$\overline{w}\underline{w} = \frac{\frac{c}{\overline{a}\underline{a}}(w - h/a)}{w - 1}.$$

Denote $\lambda := h/a$ and $\eta := c/(\overline{a}\underline{a})$. By simple manipulation, we can rewrite (A) as $\eta \equiv 1$, (B) as $\overline{\eta}\overline{\lambda} = \overline{\overline{\lambda}}\underline{\eta}$, (C) as $\overline{\overline{\eta}}\underline{\underline{\eta}} = \lambda\eta$, (A') as $\eta\lambda = \overline{\lambda}\underline{\lambda}$, (B') as $\underline{\eta}\underline{\lambda} = \overline{\eta}\underline{\underline{\lambda}}$, and (C') as $\overline{\overline{\lambda}}\underline{\underline{\lambda}}\overline{\overline{\eta}}\underline{\underline{\eta}} = \overline{\lambda}^{[3]}\underline{\lambda}_{[3]}\eta$.

Combining (A) and (A') yields for λ the equation $\overline{\lambda}\underline{\lambda} = \lambda$. This is option (1) for equation (3.7c). If (B') holds together with (A), we find that λ is period-1 (this case falls into either of the options (1) and (2)), while the combination of (A) and (C') results in

$$\overline{\lambda}^{[3]}\underline{\lambda}_{[3]} = \overline{\overline{\lambda}}\underline{\underline{\lambda}},$$

the other possibility in option (1).

An equation that satisfies (B) and (A') satisfies (A) and (B') after the change $w \to 1/y$, and thus needs no separate consideration. Combining (B) and (B') gives for λ the relation $\overline{\lambda}\underline{\lambda} = \overline{\overline{\lambda}}\underline{\underline{\lambda}}$, which can be further combined with (B) to obtain $\eta\underline{\eta} = \overline{\overline{\eta}}\underline{\eta}_{[3]}$ and thus option (2). An equation satisfying (B) and (C') can be changed into one satisfying (C) and (B') by $w \to 1/y$.

The combination of (C) and (A') gives the same result as (A) and (C') above after the change $w \to 1/y$. When (C) and (B') are both true, we find after a trivial manipulation option (3). Finally, (C) and (C') give option (4).

We still need to see what happens if either $a \equiv 0$ or $h \equiv 0$, because these facts affect the results of the iteration and the consideration of y. First, let $a \equiv 0$. We may assume that m_{j+} is positive. If $m_{j-} \neq 0$, it suffices to iterate the equation to the positive direction. Namely, we obtain an infinite sequence of poles and zeros the orders of which appear periodically, and comparing these orders we obtain a contradiction by Lemma 3.11. Observe that this iteration is done separately when $m_{j+} > k_j$ and when $m_{j+} < k_j$.

If, however, $z_j - 1$ is a zero of $w - h$, it is possible to avoid the contradiction. Namely, let $w(z_j - 1) = h(z_j - 1) + 0^{n_j}$, $n_j < k_j$, and $w(z_j - 3) = h(z_j - 3) + 0^{N_j}$. If these orders are large enough $\gg S(r,w)$, we do not get a contradiction from Lemma 3.11. Since now $w(z_j-3) = c(z_j-2)/h(z_j-1)$, this would mean that $c = \overline{h}\underline{h}$. Now transforming $w \to h/w$ we find the equation

$$\overline{w}\underline{w} = \frac{-1}{w-1}, \tag{4.24}$$

which is a degenerate special case of (3.7a), option (2).

Suppose then that $h \equiv 0$. This reasoning is very similar to the foregoing, so we will skip most of the details. Iteration with $m_{j+} > 0$ gives an sequence where w takes the consecutive values

$$a(z_j), \infty, \varphi, 0, 0, \varphi, \infty, \varphi, 0, 0, \ldots,$$

where φ denotes a nonzero finite value (possibly different at every occurence) or a zero or pole with an insignificant order compared to m_{j+}, and the zeros and poles all have the same order m_{j+} modulo a small effect from the coefficients. Now there are two ways to escape a contradiction: first, if $w(z_j+2) = a(z_j+2)$, $\gg S(r,w)$, which again means that $c = \overline{a}\underline{a}$, then making the transformation $w \to wa$ we find the equation

$$\overline{w}\underline{w} = \frac{w}{w-1},$$

which is included in options (1) and (2) for (3.7c). The second possibility is that $w(z_j+7) = a(z_j+7)$, $\gg S(r,w)$, in which case the sequence is $(a, \infty, \varphi, 0, 0, \varphi, \infty, a)$. By the result of the iteration for $w(z_j+7)$ this means

$$a\underline{c}_{[3]}\underline{a}\overline{\overline{c}} = c\underline{a}_{[4]}\underline{c}\overline{a}^{[3]},$$

which implies that $\eta = c/(\overline{a}\underline{a})$ satisfies $\overline{\overline{\eta}}\underline{\eta}_{[3]} = \eta\underline{\eta}$. This is included in option (2) for equation (3.7c).

Case $p = 0$. If $a \equiv 0$, we have just equation (3.7d). If not, we can change w into $y = 1/w$, which gives us a form where $p = q = 1$ and the coefficient in the denominator is identically zero:

$$\overline{y}\underline{y} = \frac{-\frac{a}{c}(y - 1/a)}{y}.$$

Equations of this form were considered above, and equation (4.24) was found.

Case $p = 2$. If no root is identically zero, we find that $w(z_j+3)$ is finite. If it is not equal to $a(z_j+3)$, continuing the iteration gives a sequence of the form

$$0^{m_{j+}}, \varphi, \infty^{m_{j+}}, \infty^{m_{j+}}, \varphi, 0^{m_{j+}}, \varphi, \infty^{m_{j+}}, \ldots,$$

from which a contradiction follows by Lemma 3.11 (again, the orders of the poles and zeros are given only modulo the small error). In fact, the stronger Lemma 3.12 also applies, since $\deg_w R = 2$, but the result is the same. By the form of the second iterate $w(z_j+3)$ we then have

$$\overline{\overline{a}}\underline{a} = c\overline{c}. \tag{4.25}$$

From this point on we must follow three separate routes. Recall that when we were considering the case of vanishing coefficients in equation (4.1a), we found, after the change to $y = 1/w$, equation (4.18), and promised to deal with it later. Now, if we are working with (4.18), then we have the results in Table 1 on page 27 at our disposal, as well. Equation (4.25) translates into the notation of (4.18) as

$$\frac{1}{\overline{\overline{h}}_2\underline{h}_2} = \frac{a_1 a_2 \overline{a}_1 \overline{a}_2}{c h_2 \overline{c}\overline{h}_2}.$$

Combining this with case (i) in Table 1 lets us write

$$\nu\overline{\nu}\overline{\overline{\lambda}}\underline{\lambda} = \lambda\overline{\lambda}$$

for $\lambda := h_2/a_1$ and $\overline{\nu}\underline{\nu} = 1$, $\nu \not\equiv 1$ (recall that ν, obtained in Subsection 4.2.1, is such that $a_2 = \nu a_1$). Then doing the transformation $w \to a_1 w$ in the original (4.1a) with $h_1 = 0$ results in

$$\overline{w}\underline{w} = \frac{w^2 - \lambda w}{(w-1)(w-\nu)}.$$

This is contained in option (1) for equation (3.7a) with $\mu \equiv 0$.

If (ii) is true, instead, we get

$$\overline{\overline{\lambda}}\underline{\lambda} = \lambda\overline{\lambda}, \tag{4.26}$$

and the transformation $w \to a_1 w$ gives

$$\overline{w}\underline{w} = \frac{\eta w^2 - \lambda w}{(w-1)(w-\nu)}.$$

Here η is period-2 and $\overline{\eta} = \nu$, so this result is contained in option (2) for equation (3.7a).

Finally, if $a_1 = a_2$, we find similarly the equation

$$\overline{w}\underline{w} = \frac{w^2 - \lambda w}{(w-1)^2},$$

where λ satisfies (4.26). This closes the consideration of equation (4.18).

When considering equation (4.20) similarly with a vanishing coefficient, we found equation (4.22). In this case (4.25) translates into

$$\frac{1}{\underline{h}_1\overline{\overline{h}}_1} = \frac{1}{c h_1 \overline{c}\overline{h}_1},$$

which implies that the function $\overline{h}_1\underline{h}_1/ch_1^2$ satisfies (3.13). Changing $w \to h_1/w$ in the first equation in (4.20) with $h_2 \equiv 0$ results in a special case of equation (3.7b) with $\lambda \equiv 0$.

Now, having dealt with equations (4.18) and (4.22), we look at the case which is of the form $p = 2$, $q = 1$ to begin with. In it, we have not yet changed w into $1/y$, and we will do that now, obtaining

$$\overline{y}\underline{y} = \frac{-\frac{a}{ch_1h_2}y(y-1/a)}{(y-1/h_1)(y-1/h_2)}.$$

The reasoning in Section 4.2 can now be applied to this equation in order to gain information on $1/h_1, 1/h_2$. Suppose $h_1 \neq h_2$. Either y (and thus w) satisfies a Riccati difference equation, or one of the following holds (compare to Table 1):

$$(i) \qquad -\frac{a}{\nu c h_1^2} = \frac{1}{\overline{h}_1\underline{h}_1}, \qquad \overline{\nu}\underline{\nu} = 1, \nu \not\equiv 1$$

$$(ii) \qquad -\frac{a}{ch_1h_2} = \frac{1}{\overline{h}_1\underline{h}_2} = \frac{1}{\overline{h}_2\underline{h}_1}.$$

Combining (i) with (4.25) one can deduce that the relation $\nu\underline{\lambda}\overline{\overline{\lambda}} = \underline{\nu}\lambda\overline{\lambda}$, where $\lambda = h_1/a$, holds. Thus changing $y \to y/h_1$, we have the equation

$$\overline{y}\underline{y} = \frac{y^2 - \lambda y}{(y-1)(y-\nu)},$$

which falls into option (1) for equation (3.7a) with $\mu \equiv 0$. Similarly, combining (ii) with (4.25), we find the equation

$$\overline{y}\underline{y} = \frac{\overline{\nu}y^2 - \lambda y}{(y-1)(y-\nu)},$$

where $\overline{\nu} = \underline{\nu}$, $\nu \not\equiv 1$, and λ satisfies (4.26). This is option (2) for equation (3.7a) with $\mu \equiv 0$. The case $h_1 = h_2$ yields the same equation with $\nu \equiv 1$.

If $a \equiv 0$, the change to $y = 1/w$ gives an equation of the form (4.1b), which has been treated in Section 4.2, and was seen to lead either to w satisfying a Riccati difference equation or to a contradiction.

Now, consider the case that $h_2 \equiv 0$ (the case $h_1 \equiv 0$ is obviously identical), and denote for simplicity $h = h_1$. The iteration now yields an infinite sequence of the form

$$0^{m_{j+}}, \varphi, \infty^{m_{j+}}, \infty^{m_{j+}}, \varphi, 0^{m_{j+}}, 0^{m_{j+}}, \varphi, \infty^{m_{j+}}, \ldots,$$

and from this we obtain a contradiction as before unless $w(z_j+3) = a(z_j+3)$, $>S(r,w)$. Changing w to $1/y$ gives an equation of the same form, only with different coefficients, and so repeating the same reasoning for it we find the conditions

$$\overline{\overline{a}}\underline{a} = c\overline{c}, \qquad \frac{1}{\overline{\overline{h}}\underline{h}} = \frac{a\overline{a}}{c\overline{c}h\overline{h}}.$$

Combining these gives us (4.26) for $\lambda := h/a$. From the first equation we get that $\eta := ca/(\overline{a}\underline{a})$ satisfies $\eta\overline{\eta} = 1$. If we replace λ with $\lambda\eta$, it still satisfies (4.26). Making the transformation $w \to wa$, we have equation (3.7b).

The case when both h_1, h_2 are identically zero is simpler. Similarly as above, we see that more than $S(r,w)$ sequences must be of the form

$$(\varphi, a+0^{m_{j-}}, \infty^{m_{j-}}, \infty^{m_{j-}}, a+0^{k_j}, \infty^{m_{j+}}, \infty^{m_{j+}}, a+0^{m_{j+}}, \varphi),$$

which implies again that η, defined as above, solves $\eta\overline{\eta} = 1$. The transformation $w \to wa$ gives

$$\overline{w}\underline{w} = \frac{\eta w^2}{w-1},$$

a special case of (3.7b).

We have now treated all the possible degrees of P and Q in equation (3.5) with all the possible subcases, and so the proof of Theorem 3.4 is completed.

5. The Family of Difference Painlevé V

Throughout this chapter we assume that w is an admissible meromorphic solution of equation (3.8), and $\rho_2(w) < 1$. The treatment here is very similar to the proof of Theorem 3.4. Because of this, we will not repeat many of the identical reasonings explicitly. It is recommended that the reader is familiar with Chapter 4 before continuing.

5.1. **Preliminaries.** We will first restrict the degrees p and q of the polynomials in (3.8). This reasoning is essentially due to Grammaticos, Tamizhmani, Ramani, and Tamizhmani in [11], who considered equations with constant coefficients.

When the coefficients are nonconstant, Lemma 3.13 needs to be applied similarly as in the previous chapter. Given $\epsilon > 0$, there are at most $S(r, w)$ points z_j where $Q(z_j, w) = 0^{k_j}$, but where $L(w)$ has a pole of order greater than $(1 + \epsilon)k_j$ or less than $(1 - \epsilon)k_j$ due to poles or zeros of $P(z_j, w)$. The combined effect of all such points can be included in the error term, and so we only consider the rest of the zeros of $Q(z, w)$ in all what follows.

Equation (3.8) can be written as

$$\overline{w}\underline{w} - \overline{w}/w - \underline{w}/w = \frac{P(z,w) - Q(z,w)}{w^2 Q(z,w)} =: K(z,w). \tag{5.1}$$

Applying the fact that $T(r, \overline{w}) \leq (1 + \epsilon)T(r + 1, w) + S(r, w)$ [1, Lemma 1], and Theorems 3.10 and 3.9 with $f = \underline{w}$, $h = 1/w$ and $g = \overline{w}$ gives

$$\begin{aligned}(\deg_w K)T(r,w) &\leq T(r,\overline{w}) + T\left(r, \frac{1}{w}\right) + T(r,\overline{w}) + S(r,w) \\ &\leq (3 + 2\varepsilon)T(r,w) + S(r,w),\end{aligned}$$

so that $\deg_w K \leq 3$. It might happen that $P - Q$ has w as a factor, although P and Q share no factors by assumption. Obviously $q < 4$ in any case. If $q \leq 2$ and $P - Q = wT$, where T is at most cubic, then $p \leq 4$.

We show that $q = 3$ is impossible. First, observe that w does not divide Q, since otherwise w would also divide $P = Q + w^2S$, where S is at most quadratic, but P and Q have no common factors. Thus, when $w = 0$, Q can have a zero only due to the coefficients, and there are at most $S(r, w)$ such zeros. The right hand side of

(5.1), S/Q, has poles essentially only in the zeros of Q (again, the coefficients might give a small amount of additional poles). This implies that

$$N\left(r, \frac{S}{Q}\right) \leq N(r, \overline{w}\underline{w}) + S(r, w).$$

By Theorem 3.5 and (5.1),

$$m\left(r, \frac{S}{Q}\right) = m(r, \overline{w}\underline{w}) + S(r, w),$$

and so $T(r, S/Q) \leq T(r, \overline{w}\underline{w}) + S(r, w)$. Thus we have

$$\begin{aligned} 3T(r, w) &= T\left(r, \frac{S}{Q}\right) + S(r, w) \leq T(r, \overline{w}\underline{w}) + S(r, w) \\ &\leq T(r, \overline{w}) + T(r, \underline{w}) + S(r, w) \leq 2(1+\epsilon)T(r, w) + S(r, w), \end{aligned}$$

a contradiction. To summarise, we have the limitations

$$q \leq 2, \quad p \leq 4,$$

and if $p = 4$ or $q = 2$, then Q and P must have the same zero-order term.

Next, we prove an analogue of Lemma 4.1:

Lemma 5.1. *Whenever $q \geq 1$ in (3.8), $m(r, w) = S(r, w)$.*

Proof. Suppose w is transcendental. We can rearrange (3.8) into

$$\underbrace{w^{q-1}L(w)}_{=:H(z,w)}\, w = \Psi(z, w),$$

where Ψ is a difference polynomial in w of degree at most $4 + (q-1)$. Applying now Theorem 3.8 implies that $m(r, w) = S(r, w)$.

Next, suppose w is rational, which implies that the coefficients in R are constants. Denote $d := \deg w$, and let N be the number of poles of w. We aim to prove that $N = d$. The degree of $L(w)$ is $\max\{p, q\} \cdot d$.

Suppose $p < q$. Then the number of zeros of $L(w)$ is dq by its degree and $pd + (q-p)N$ by the form of the right hand side, which implies that $N = d$.

Next, let $p = q$. Then $L(w)$ has the same amount of zeros and poles, dq, and thus the same must be true for w, i.e. $N = d$.

Finally, if $p > q$, $L(w)$ has degree pd, and exactly $dq + (p-q)N$ poles. Suppose that k terms are cancelled when we write out the factored rational expression for $L(w)$, so that the number of poles is $4N - k$. Then $4d - k = dp$ so that $k = d(4-p)$. The number of poles is $4N - k = dq + (p-q)N$, which again implies that $N = d$. □

Lemma 5.2. *Suppose that $q = 2$ or $p > 2$. Then P and Q in (3.8) have the same zero-order term.*

Proof. We already know this fact when $p = 4$ or $q = 2$, so suppose that $p = 3$ and $q < 2$. If $w \equiv 0$ is a solution to (3.8), then either we will directly obtain the desired result by substitution (if none of the roots vanishes identically), or find a contradiction (if zero is a root). If $w \equiv 0$ is not a solution, then we can again use the modified version of [14, Corollary 3.4] as in the proof of Lemma 4.2 to conclude that $m(r, 1/w) = S(r, w)$, which means that w has a large number of zeros.

Consider points z_j such that $w(z_j) = 0^{k_j}$, and $w(z_j \pm 1) = \infty^{m_{j\pm}}$. If there are more than $S(r, w)$ points z_j such that $m_{j\pm} < Ck_j$ for both choices of $\pm$ and some $C < 1$, we again find either a contradiction or the desired result. Hence we suppose that m_{j+}, say, is at least $(1-\epsilon)k_j$ for all except at most $S(r, w)$ points.

When $q = 0$, both $w(z_j + 2)$ and $w(z_j + 3)$ are infinite with orders roughly $2m_{j+}$ and m_{j+}, respectively. When $q = 1$, w takes at $z_j, z_j + 1, \ldots$ the consecutive values

$$0, \infty, \infty, 0, \infty, \ldots,$$

and all these poles and zeros have roughly the same order. Hence we have that

$$N\left(r, \frac{1}{w}\right) \leq \frac{1}{2}N(r + 2, w) + S(r, w),$$

which gives a contradiction by Theorem 3.10, since $m(r, 1/w) = S(r, w)$. These contradictions prove the claim. □

Corollary 5.3. *Suppose $q = 2$ or $p > 2$. Then none of the roots of P or Q in (3.8) vanish identically.*

Proof. By Lemma 5.2, if one root of P is zero, one root of Q must also be zero, and vice versa, contradicting the fact that P and Q have no common factors. □

We will not refer to Corollary 5.3 explicitly, but simply do not consider the cases of vanishing coefficients whenever $q = 2$ or $p > 2$. In fact, we will shortly see that $q = 2$ implies $p = 4$, so that the condition "$q = 2$ or $p > 2$" can be simplified into $p > 2$.

The case of $p = q = 0$ has just a small function in the place of $R(z, w)$, and we will not consider it further. In the other cases we will factor P in (3.8) into $c(w - h_1) \cdots (w - h_p)$, where h_n are algebroid functions, and Q into $(w - a_1)(w - a_2)$ or $w - a$ when $q = 2$ or $q = 1$, respectively.

We are not proving a pre-formulated theorem like we were in Chapter 4, but our aim is similar: to find restrictions on the coefficients in $R(z, w)$. In the following two sections, we derive such restrictions, and all of our findings are then summarised in Section 5.4. Even though we do not state a definite theorem, all possible subcases are handled below, and in that sense the results given in Section 5.4 are just as complete as those in Theorem 3.4.

5.2. **Points where Q vanishes.**

Equations with $q = 2$. We can prove, exactly as in Lemma 4.2, that $w - a_m$ has a large number of zeros in the sense that

$$N\left(r, \frac{1}{w - a_m}\right) = T(r, w) + S(r, w).$$

First, let $p < 4$. We show that this case gives a contradiction. Assume $a_1 \neq a_2$, and consider the points z_j where $w(z_j) = a_m(z_j) + 0^{k_j}$. Discarding at most $S(r,w)$ of these points if necessary, we can say that either or both of $w(z_j+1)$ and $w(z_j-1)$ are infinite, and the sum of the orders of these poles is equal to the order k_j modulo a small error term. By (3.8), the next value of w following a pole at $z_j \pm 2$ will be a zero, regardless of the value of $p < 4$. Thus, all except at most $S(r,w)$ zeros of $w - a_m$ are contained in sequences of one of the three forms

$$\begin{gathered}(0, \infty^{m_{j-}}, a_m + 0^{k_j}, \infty^{m_{j+}}, 0),\\ (\varphi, a_m + 0^{k_j}, \infty^{k_j}, 0),\\ (0, \infty^{k_j}, a_m + 0^{k_j}, \varphi),\end{gathered}$$

where we do not care about the orders of the zeros and have only given the other orders modulo the possible small error. Because this holds separately for both a_1 and a_2, and because $N(r, 1/(w - a_m)) = T(r,w) + S(r,w)$ for both choices of m, we have arrived to a situation where $N(r+1, w) \geq 2T(r,w) + S(r,w)$, which is obviously a contradiction. This reasoning easily extends to the case that $a_1 = a_2$, as well (then the above sequences will contain roughly twice more poles compared to the zeros of $w - a_1$).

Hence we must have $p = 4$, so that $\deg_w R = 4$. Recall the definition of a sequence $L(z_j, w)$ from Section 4.2. We suppose that, $\gg S(r,w)$ z_j, the sequence $L(z_j, w)$ contains only one zero of $w - a_m$. Let this zero be of order k_j. Then the pole of $L(w)$ at z_j is of order m_j, where $m_j > (1-\epsilon)k_j$. Hence, there are more than $S(r,w)$ sequences $L(z_j, w)$ such that

$$N\left(r, \frac{1}{Q(z,w)}\right) \leq \beta N(r+1, w),$$

where $\beta < 2$ (compare to the reasoning in Section 4.2). Since $p > q$, $R(z,w)$ has poles both when $Q(z,w) = 0$ and when w itself has a pole. We have, in fact,

$$N(r,R) = N(r, L(w)) = 2N(r,w) + N\left(r, \frac{1}{Q(z,w)}\right) + S(r,w). \tag{5.2}$$

Now Lemma 3.12 yields a contradiction. Thus, there must be at least two zeros of $w - a_m$ in all except at most $S(r,w)$ sequences $L(z_j, w)$. Again, one of the possibilities (i)–(iv) on page 25 is valid. This reasoning can be repeated with slight modifications also in the case that $a_1 = a_2$, and then condition (i) is the only possibility.

The equation we are considering is now

$$(\overline{w}w - 1)(\underline{w}w - 1) = \frac{c(w-h_1)(w-h_2)(w-h_3)(w-h_4)}{(w-a_1)(w-a_2)}. \tag{5.3}$$

We can show, very much like in Section 4.2, that w satisfies a Riccati equation in the cases (iii) and (iv). The only difference is in showing that w has at most $S(r,w)$ poles outside of the sequences $L(z_j, w)$. Suppose, on the contrary, that the integrated counting function counting only such poles is $N^*(r,w) \geq CT(r,w)$ for some $C > 0$ in a set of infinite logarithmic measure. By equation (5.3), $N^*(r, L(w)) = 2N^*(r, w) + S(r, w)$. By Theorem 3.6,

$$\begin{aligned} 4T(r,w) &= (N(r,L(w)) - N^*(r,L(w)) + N^*(r,L(w)) + S(r,w) \\ &\leq 4(N(r+1,w) - N^*(r+1,w)) + 2N^*(r+1,w) + S(r,w) \\ &\leq (4-2C)T(r+1,w) + S(r,w), \end{aligned}$$

which implies that $\rho_2(w) \geq 1$ by Theorem 3.10. Hence w has only a small amount of poles outside of the sequences $L(z_j, w)$. Now that this has been shown, we can proceed exactly as on page 25, except that to finally get to the contradiction, we must also use (5.2). By that reasoning, we conclude that whenever (iii) or (iv) is true, w satisfies a Riccati difference equation (4.7).

Observe that the above reasoning about the poles of w is not dependent on (iii) or (iv) being true, but only on the equation being of form (5.3). This means that *in any case w has at most $S(r, w)$ poles that are not in sequences $L(z_j, w)$.* This is an important fact in the below treatment of another kind of singularities. Before going into that, we briefly look at what happens in cases (i) and (ii).

Suppose that (i) holds and $a_1 \neq a_2$. Then (4.4) holds $\gtrdot S(r, w)$, and starting from the assumption that $w(z_j) = a_1(z_j)$, we get by (5.3) that

$$w(z_j + 2) = a_1(z_j + 2) = \frac{c(z_j + 1)}{a_1(z_j)}, \quad \gtrdot S(r, w).$$

This implies that $c = \underline{a}_1\overline{a}_1$. Similarly, by starting from $w(z_j) = a_2(z_j)$, we get that $c = \underline{a}_2\overline{a}_2$. Compare this to the treatment of cases (i) and (ii) on page 25. Although the considered equation is more complicated, we obtain the same results as for equation (4.1a) in Section 4.2: in case (ii) we get

$$c = \overline{a}_1\underline{a}_2 = \overline{a}_2\underline{a}_1.$$

This result is obtained as above, using (5.3) together with the condition (ii). In the case that $a_1 = a_2 =: a$ we find $c = \overline{a}\underline{a}$.

Equations with $q = 1$. Now we write $Q(z, w) = w - a$. Suppose that $w(z_j) = a(z_j) + 0^{k_j}$ and $w(z_j + 1) = \infty^{m_j}$. (Observe that the equation is symmetrical so that iterating in the other direction gives eventually the same results; only with shifts up changed to shifts down and vice versa.) We may suppose that m_j/k_j is bounded from below by a positive constant. In what follows, we compute a few of the following iterates for different values of p. Some of the more complicated computations were performed by the aid of computer, and the details are largely omitted.

Let first $p = 4$. Then,

$$N(r, L(w)) = 3N(r, w) + N\left(r, \frac{1}{Q(z, w)}\right) + S(r, w).$$

Now $w(z_j+2)$ is infinite, while

$$w(z_j+3) = \frac{c(z_j+2)c(z_j+1)}{a(z_j)}.$$

Following the same reasoning as in Chapter 4, by Lemma 3.11 we must restrict this to be equal to $a(z_j+3)$ or otherwise w will have hyper-order at least one. Since this must hold $\gtrdot S(r,w)$, we find

$$\overline{\overline{a}}\underline{a} = \overline{c}c.$$

Next, consider $p = 3$. Now, $w(z_j+2) = c(z_j+1)/a(z_j)$, and $w(z_j+3) = a(z_j)/c(z_j+1)$, provided that $w(z_j+2) \neq a(z_j+2)$, and $w(z_j+4)$ will again be infinite, provided that $w(z_j+3) \neq h_n(z_j+3)$, for any $n = 1,2,3$. Suppose that there are more than $S(r,w)$ sequences of this type. Then, comparing the orders of the points (all of them are of order m_j modulo a small error), we see that Lemma 3.11 will yield a contradiction.

There are two possibilities to avoid the contradiction: first, that more than $S(r,w)$ sequences are of the form $(a(z_j), \infty, a(z_j+2), \varphi)$ for some finite value φ, which implies that $c = \overline{a}\underline{a}$, and second, that we have a large number of sequences of the form $(a(z_j), \infty, \varphi, h_n(z_j+3))$. This second possibility in fact leads to a contradiction, as will be seen later.

In the case that $p < 2$ we get a sequence of alternating zeros and poles for $w(z_j+2)$, $w(z_j+3)$, ..., leading to a contradiction via Theorem 3.10 similarly as on several occasions before. This happens also if one of the coefficients a and h_1 vanishes identically. Hence the cases that $q = 1$ and $p < 2$ can be ignored.

Finally, let $p = 2$, and suppose that the coefficients do not vanish identically. Again, $w(z_j+2) = 0$, and $w(z_j+3)$ will be infinite unless

$$c(z_j+2)h_1(z_j+2)h_2(z_j+2)a(z_j) - a(z_j+2)c(z_j+1)$$

vanishes. The next iterate is

$$w(z_j+4) = -\frac{c(z_j+1)a(z_j+2)c(z_j+3)}{a(z_j)c(z_j+2)h_1(z_j+2)h_2(z_j+2)}.$$

We get a contradiction via Lemma 3.11 unless either

$$\overline{c}\overline{h}_1\overline{h}_2\underline{a} = \overline{a}c,$$

or z_j+4 is another zero of $w-a$, i.e.

$$-\underline{c}a\overline{c} = \overline{\overline{a}}ch_1h_2\underline{\underline{a}}.$$

In the case that some of the coefficients a, h_1 and h_2 vanish identically, the pole at z_j+3 cannot be avoided and we always end up in a contradiction.

5.3. **Points where $w\underline{w}-1$ vanishes.** Assume $p > 0$. We aim to prove that for each $n = 1, \dots, p$,

$$h_n\overline{h}_k = 1, \text{ or } h_n\underline{h}_k = 1. \tag{5.4}$$

for some $k = 1, \dots, p$ (it is also possible that $n = k$). Following the same reasoning as in the proof of Lemma 4.2, we see that

$$N\left(r, \frac{1}{w - h_n}\right) = T(r, w) + S(r, w),$$

as long as h_n is not a solution of (3.8), i.e. if h_n does not satisfy $h_n\overline{h}_n = 1$. Thus, we may assume that there is a large number of zeros of $w - h_n$; otherwise (5.4) holds as desired.

Consider now the points z_j where $w(z_j) = h_n(z_j)$. From equation (3.8) we see that

$$w(z_j)w(z_j \pm 1) = 1,$$

for all except at most $S(r, w)$ points z_j and for either or both choices of $\pm$. In principle, $z_j \pm 2$ can be poles of w. To avoid this kind of a pole we must have $w(z_j \pm 1) = 1/h_n(z_j) = h_k(z_j \pm 1)$ for some $k = 1, \dots, p$. If this holds $\gg S(r, w)$ we have exactly (5.4).

It turns out that there are some cases where (5.4) does not hold for one value of n. These exceptions occur when $q = 0$ and when $q = 1$, $p = 3$. In all the other cases, however, we will prove that (5.4) holds for all n.

In what follows, we fix $n = 1, \dots, p$, and suppose that (5.4) is not satisfied. This means that, as $w(z_j) = h_n(z_j)$, we have $w(z_j \pm 1) = 1/h_n(z_j)$, and $w(z_j \pm 2) = \infty$ for either or both choices of the $\pm$ sign, for all except at most $S(r, w)$ such points z_j. Then all except at most $S(r, w)$ zeros of $w - h_n$ are in these kind of sequences. This means, in particular, that all except at most $S(r, w)$ poles of w are in these sequences, as well, since there are roughly the same amount of poles of w and zeros of $w - h_n$ in these sequences, and both $N(r, w)$ and $N(r, 1/(w - h_n))$ are equal to $T(r, w) + S(r, w)$.

We make a few more general observations before looking at individual cases. It is impossible to have a triple or quadruple factor $w - h_n$ in P. That would imply that the right hand side of (3.8) has a zero of order (roughly) $3k_j$ or $4k_j$, respectively, at a point where $w(z_j) = h_n(z_j) + 0^{k_j}$, but the left hand side cannot have a zero of this high order. If h_n appears *twice* in P, it is possible to have $w(z_j) = h_n(z_j) + 0^{k_j}$ and essentially $w(z_j \pm 1) = 1/h_n(z_j) + 0^{k_j}$ for *both* choices of $\pm$, but this is the only possibility.

Secondly, observe that a sequence where w takes the values $(\infty, h_n(z_j), 1/h_n(z_j))$ cannot occur for more than at most $S(r, w)$ points; this follows from (3.8) by considering the orders of the points.

Lastly, we often find that $w(z_j \pm 1) = 1/h_n(z_j)$ would have to be equal to $a_m(z_j \pm 1)$, $m = 1, 2$, $\gg S(r, w)$, i.e. that $h_n\overline{a}_m = 1$ or $h_n\underline{a}_m = 1$. This however, is not possible if we still assume $w(z_j \pm 2)$ to be a pole, as can be seen from equation (3.8). Namely then we have, for example in the case that $q = 1$, $p = 2$,

$$\begin{aligned}(w(z_j+2)w(z_j+1)-1)(w(z_j)w(z_j+1)-1)\\ = \frac{c(z_j+1)(w(z_j+1)-h_k(z_j+1))(w(z_j+1)-h_n(z_j+1))}{w(z_j+1)-a(z_j+1)}.\end{aligned}$$

Here the left hand side is finite or a pole with a small order, since $w(z_j+2)=\infty$ and $w(z_j) = 1/w(z_j+1)$, both with roughly the same order. The right hand side, however, has a pole, since $w(z_j+1) = 1/h_n(z_j) = a(z_j+1)$, but because P and Q have no common factors the numerator cannot be zero (with a significant order). This contradiction can be obtained exactly in the same way for other values of q and p (here we just took the smallest values for notational simplicity).

Case $q = 2$. We may assume that condition (i) or (ii) is true (if not, w satisfies a Riccati difference equation).

Most poles of w must be contained in some sequence $L(z_j, w)$ containing at least two distinct zeros of $w - a_m$. Because $w(z_j+1) = 1/h_n(z_j)$ cannot be equal to $a_m(z_j+1)$ for many points, the next iterate after the pole at z_j+2 must be a zero of $w - a_m$. Using (3.8), we get that $w(z_j+3) = c(z_j+2)h_n(z_j)$. Restricting this to be equal to $a_m(z_j+3)$, $\gtrdot S(r,w)$, we get $c\underline{h}_n = \overline{a}_m$, and combining this with the representation for c from (i) or (ii) gives $h_n\overline{a}_m = 1$, which is impossible.

Case $q = 1$. First, let $p = 4$. Computing the values of w at $z_j \pm 2, 3, 4$, we find generally a sequence of the form

$$h_n(z_j) + 0^{k_j}, 1/h_n(z_j) + 0^{m_j}, \infty^{m_j}, \infty^{m_j}, \varphi,$$

where φ is some finite value or a zero or pole with an insignificant order, and the orders are given only modulo a small error (if there is no zero of $w - 1/h_n$ at $z_j - 1$, or if h_n appears twice in P, the orders m_j and k_j are roughly equal, while otherwise $m_j < k_j$, but we get another similar sequence when iterating in the other direction). By our assumption, all except at most $S(r,w)$ zeros of $w - h_n$ are in these kind of sequences. This means that most poles of w are present here, as well, but we saw in the previous section that there are more than $S(r,w)$ sequences of the form $(a(z_j), \infty, \infty, a(z_j+3))$, and this is again a contradiction, as it would imply that $h_n\overline{a} = 1$ or $h_n\underline{a} = 1$.

Next, suppose $p = 3$. Now w takes at $z_j + 0, 1, 2, 3$ the values

$$h_n(z_j) + 0^{k_j}, 1/h_n(z_j) + 0^{m_j}, \infty^{m_j}, \varphi, \tag{5.5}$$

where $\varphi = w(z_j+3) = c(z_j+2)h_n(z_j)$. By our assumption, all except at most $S(r,w)$ zeros of $w - h_n$ are in these kind of sequences, implying that in fact all except at most $S(r,w)$ poles of w are in these sequences, as well. We saw earlier, when considering the zeros of Q, that more than $S(r,w)$ poles must appear in sequences of the form $(a(z_j), \infty, a(z_j+2), \varphi)$ or $(a(z_j), \infty, \varphi, h_n(z_j+3))$. Thus for all except at most $S(r,w)$ sequences, either of these possibilities must coincide with the sequences (5.5).

The first possibility implies that $1/h_n = \overline{a}$, which is impossible. In the second case we have either the same impossible result or $c = \overline{a}\underline{\underline{h}}_n$. This is the exception we must allow to (5.4) in the case $q = 1$, $p = 3$.

Since $p < 2$ is impossible as seen earlier, the only case left is $p = 2$. It was also observed earlier that the coefficients cannot vanish identically in this case. We know that there are more than $S(r, w)$ sequences of the form $(a, \infty, 0)$, and iterating the equation with the current assumptions gives a sequence of the form $(h_n, 1/h_n, \infty, 0)$. By assumption, the latter sequences contain all except at most $S(r, w)$ poles of w. To avoid a contradiction, these sequences must coincide so that $\underline{h}_n a = 1$, which is impossible.

Case $q = 0$. We have not shown that w would have (a large number of) poles when $q = 0$, but if it does not, then clearly (5.4) must be true.

If $p = 4$, iteration gives a pole at every following point $z_j + 2, 3, \ldots$, which is clearly a contradiction. The reasoning is not essentially changed if we suppose that h_n is a double factor in P.

Let then $p = 3$ and $h_1 \neq h_2 \neq h_3 \neq h_1$. Then w takes at $z_j + 0, 1, 2, 3, 4, 5$ the values

$$h_n(z_j), \frac{1}{h_n(z_j)}, \infty, \infty, \varphi, \frac{1}{\varphi},$$

where $\varphi = c(z_j + 3)c(z_j + 2)h_n(z_j)$. To avoid a contradiction by Theorem 3.10, we must restrict φ to be $1/h_n(z_j + 3)$, and thus $\overline{c}c\underline{\underline{h}}_n\overline{h}_n^{[3]} = 1$. This is an alternative for (5.4). In the case that $h_1 = h_2$, say, a similar sequence is impossible, and we only get a contradiction, since

$$N\left(r, \frac{1}{w - h_n}\right) \leq 2N(r + 3, w) + S(r, w).$$

When $p = 2$ and $h_1 \neq h_2$, we find $w(z_j + 3) = c(z_j + 2)h_n(z_j)$ and $w(z_j + 4) = 1/w(z_j + 3)$. A contradiction can be avoided if the poles of w appear in a sequence of the form

$$h_n(z_j), \frac{1}{h_n(z_j)}, \infty, \frac{1}{h_k(z_j + 3)}, h_k(z_j + 3),$$

where $k \neq n$. This implies that $\overline{\overline{h}}_n c\underline{\underline{h}}_k \equiv 1$. If $h_1 = h_2$, this kind of exception is not possible.

In the case $p = 1$ we find that $w(z_j + 3) = 0$ and $w(z_j + 4) = \infty$, unless $c\underline{\underline{h}} = \overline{c}\overline{h}$. If this condition is not satisfied, there seems to be another possibility as well, namely that most poles would be in sequences of the form

$$h(z_j), \frac{1}{h(z_j)}, \infty, 0, \infty, \frac{1}{h(z_j + 6)}, h(z_j + 6), \tag{5.6}$$

where all the points have roughly the same order. This would mean (computing from the equation again), that $\overline{h}^{[3]}\underline{h}_{[3]}\overline{c}\underline{c} = -ch$. Now we can reason again similarly as in the proof of Lemma 5.2. If $w \equiv 0$ is a solution, then $-ch \equiv 1$, while if $w \equiv 0$ is not a solution, then $m(r, 1/w) = S(r, w)$, which means that there are more than $S(r, w)$ zeros of w somewhere outside of the sequences (5.6). Moreover, more than $S(r, w)$ of these zeros cannot be surrounded by poles of w, because all except at most $S(r, w)$ poles are already accounted for in the sequences (5.6). Hence, in any case, $-ch \equiv 1$.

Summary and consequences of equation (5.4). Denote an arbitrary period-k function by p_k. We know that each coefficient h_n of P satisfies $h_n\overline{h}_k = 1$ or $h_n\underline{h}_k = 1$ for some k, unless some of the exceptional cases (when $q = 0$ or $q = 1$, $p = 3$) are valid. If a function h_n satisfies $h_n\underline{h}_n = 1$, then h_n is a solution of (3.13). If two functions satisfy $h_1\underline{h}_2 = h_2\underline{h}_1 = 1$, we have that $h_1 = p_2$ and $h_2 = 1/\underline{p}_2$. A "cycle" of three functions, i.e. $h_1\underline{h}_2 = 1$, $h_2\underline{h}_3 = 1$, and $h_3\underline{h}_1 = 1$, implies that h_1 satisfies $h_1\overline{h}_1^{[3]} = 1$, and $h_2 = \underline{\underline{h}}_1$, $h_3 = 1/\underline{h}_1$. Finally, a "cycle" of four functions implies that all the functions h_n are period-4 and can be written as p_4, $1/\overline{p}_4$, $\overline{\overline{p}}_4$, and $1/\overline{p}_4^{[3]}$.

In addition to the mentioned "cycles", there are possibilities that cannot be directly solved any further. One is that we have $h_1\overline{h}_2 = 1$, but not $h_2\overline{h}_1 = 1$. This gives $h_1 = 1/\overline{h}_2$, but h_2 remains arbitrary. Another possibility arises when $p = 4$, two functions h_n are equal, say $h_1 = h_2$, and we have just $h_1\overline{h}_3 = 1$, $h_1\underline{h}_4 = 1$, which again leaves one arbitrary function. Observe that the exceptions allowed when $q = 0$ or $q = 1$, $p = 3$ are impossible for double factors.

Suppose that none of the exceptions of case $q = 0$ is valid. Then, combining the above results gives essentially six different possibilities with $p = 4$:

4a: Each h_n is a solution of (3.13).
4b: h_1, h_2 are solutions of (3.13), and $h_4 = 1/\underline{h}_3$ with h_3 arbitrary.
4c: The functions are h_1, $1/\overline{h}_1$, h_3, and $1/\overline{h}_3$, with h_1, h_3 arbitrary.
4d: h_1 solves (3.13), while h_2 solves $f\overline{f}^{[3]} = 1$, $h_3 = \underline{\underline{h}}_2$, and $h_4 = 1/\underline{h}_2$.
4e: The functions are p_4, $1/\overline{p}_4$, $\overline{\overline{p}}_4$, and $1/\overline{p}_4^{[3]}$.
4f: $h_1 = h_2$ is arbitrary, and $h_3 = 1/\overline{h}_1$, $h_4 = 1/\underline{h}_1$.

The requirement that no three functions h_n can be equal is assumed in all of these cases. One or two pairs of equal functions are possible in cases 4a–4c. As mentioned above, in certain special cases instead of arbitrary functions we have period-2 functions.

With $p = 3$ we get the following options (assuming the exceptions do not hold):

3a: Each h_n is a solution of (3.13).
3b: h_1 solves (3.13), while $h_3 = 1/\underline{h}_2$ with h_2 arbitrary.
3c: h_1 solves $f\overline{f}^{[3]} = 1$, $h_2 = \underline{\underline{h}}_1$, and $h_3 = 1/\underline{h}_1$.

When $p = 1, 2$ we have only three possibilities:

2a: Both h_1 and h_2 satisfy (3.13).
2b: $h_2 = 1/\underline{h}_1$ with h_1 arbitrary.
1a: The single function h must solve (3.13).

5.4. **Restriction of the coefficients.** In this section we collect our findings for the various cases. We have shown that, assuming w to have hyper-order less than one, equation (3.8) simplifies into

$$(\overline{w}w-1)(\underline{w}w-1)=\frac{c\prod_{n=1}^{p}(w-h_n)}{\prod_{m=1}^{q}(w-a_m)},\quad p\leq 4, q\leq 2,$$

where $c\in\mathcal{S}(w)$, and the functions h_n, a_m are small, in general algebroid, functions. We shall denote

$$H:=\prod_{n=1}^{p}h_n,$$

and drop the subscripts from a and h when $q=1$ and $p=1$, respectively, like we have done above.

Solving the coefficients from the relations we have found turns out to be somewhat easier if we assume that $H\overline{H}=1$, which is true whenever all of the functions h_n satisfy one of the conditions 4a–1a in the previous section without exceptions, and the arbitrary functions appearing in these conditions are period-2. The outcomes with arbitrary coefficients are $H\overline{H}=\overline{h}_1/\underline{h}_1$ (from 4b, 3b, or 2b), or $H\overline{H}=\overline{h}_1\overline{h}_2/(\underline{h}_1\underline{h}_2)$ (from 4c), or $H\overline{H}=h_1\overline{h}_1/(\underline{h}_1\overline{\overline{h}}_1)$ (from 4f).

We will also see what happens when all the h_n are required to be period-1, which corresponds to them being constants in the completely discrete setting. With $p=2,4$ this implies that $H\equiv\pm1$. When $p=3$ we have also $H\equiv\pm1$, except in the exceptional case when $q=1$, in which $H=h_1$ is period-1. When $p=1$ we just have $h\equiv\pm1$. Observe that the case $H\equiv-1$ when $p>1$ is possible only if two of the coefficients h_j are constants 1 and -1.

The value pairs of p and q in the following subtitles are the only possible ones. The first paragraph after each subtitle gives all the constraints we have found on the coefficients when assuming nothing extra. We only give these constraints in the form obtained when iterating the equation in the positive direction; symmetrical alternatives, where each positive shift is changed into a negative shift of the same magnitude and vice versa, are also possible.

After the first paragraph, we give some example solutions for the coefficients. In our examples, p_k and q_k are period-k functions, $\lambda\in\mathbb{C}$, and u_k is a kth root of unity. For a more general discussion on solving some of the equations, see Section 3.3.1.

Case $p=4$, $q=2$. Either w satisfies a Riccati difference equation, or one of the following is true:

$$c=\overline{a}_1\underline{a}_1,$$
$$c=\overline{a}_1\underline{a}_2=\overline{a}_2\underline{a}_1,$$
$$a_1=a_2=:a,\quad c=\overline{a}\underline{a}.$$

The functions h_n satisfy one of 4a–4e. Moreover, by Lemma 5.2, $cH=a_1a_2$.

Suppose that $H\overline{H}=1$ is true. From the first of the above possibilities we get $a_1\overline{a}_1=\underline{a}_1\overline{\overline{a}}_1$. If $c=\overline{a}_1\underline{a}_2=\overline{a}_2\underline{a}_1$, using either of these representations for c suitably, we find that both a_1 and a_2 satisfy the same equation $f\overline{f} = \underline{f}\overline{\overline{f}}$. In the last case we have $c = \overline{a}\underline{a}$ and $cH = a^2$, and we get again this same equation for a.

If all the functions h_n are period-1 so that $H \equiv \pm 1$, our reasoning simplifies remarkably. In the case that $a_1 = -a_2$ we find the equation $\overline{a}_1\underline{a}_1 = \mp a_1^2$ If we choose $H = +1$, we eventually get a contradiction by Lemma 5.2. Hence, we can take for instance $a_1 = \lambda^z p_1 u_2^{z^2}$, and $c = \lambda^{2z} p_1^2$.

In the second case, when $c = \overline{a}_1\underline{a}_2 = \overline{a}_2\underline{a}_1$, we have for c the equation $\overline{c}\underline{c} = c^2$, so that we can take for instance $c = g_1\lambda^{2z}$, and then the possible combinations for a_1 and a_2 are $a_1 = f\lambda^z$, $a_2 = h\lambda^z$, where (f, h) is is either $(p_2, \overline{p}_2)$ or (p_1, q_1). By Lemma 5.2, only $H = +1$ comes into question here.

When $a_1 = a_2 =: a$, we get for a the same solutions as above (a satisfies $\overline{a}\underline{a} = \pm a^2$), and this gives also for c the same two possible solutions. Lemma 5.2 will again remove one of the two possible equations.

To summarise, we have found the difference Painlevé V equation (3.20a), equation (3.20b), and also the variants

$$(w\overline{w} - 1)(w\underline{w} - 1) = \frac{p_2\overline{p}_2\lambda^{2z}(w - p_1)(w - 1/p_1)(w - q_1)(w - 1/q_1)}{(w - p_2\lambda^z)(w - \overline{p}_2\lambda^z)},$$

$$(w\overline{w} - 1)(w\underline{w} - 1) = \frac{\lambda^{2z}p_1^2(w - p_1)(w - 1/p_1)(w - q_1)(w - 1/q_1)}{\left(w - p_1\lambda^z u_2^{z^2}\right)^2}.$$

We stress that these are only a few examples from a large family of possible equations.

Case $p = 4$, $q = 1$. We have $\overline{\overline{a}}\underline{a} = \overline{c}c$, and h_n satisfy one of 4a–4e, and $cH = -a$ by Lemma 5.2.

If $H\overline{H} = 1$, we can combine these to find first that $a\overline{a} = c\overline{c}$, and further that a satisfies $\overline{\overline{a}}\underline{a} = \overline{a}a$.

If the h_n are period-1, so that $H \equiv \pm 1$, we have simply $c = \mp a$. The equation for a is solved by $a = p_2\lambda^z u_4^{z^2}$ for instance. The following are examples of equations obtained in this case:

$$(w\overline{w} - 1)(w\underline{w} - 1) = \frac{p_2\lambda^z u_4^{z^2}(w - p_1)(w - 1/p_1)(w - q_1)(w - 1/q_1)}{w - p_2\lambda^z u_4^{z^2}},$$

$$(w\overline{w} - 1)(w\underline{w} - 1) = \frac{p_2\lambda^z u_4^{z^2}(w - p_1)(w - 1/p_1)(w^2 - 1)}{w - p_2\lambda^z u_4^{z^2}}.$$

Case $p = 3$, $q = 1$. Either $c = \overline{a}\underline{a}$ and the h_n satisfy one of 3a–3c, or $c = \overline{a}\underline{\underline{h}}_3$ and h_1, h_2 satisfy one of 2a–2b. In both cases $cH = -a$ must hold by Lemma 5.2.

Suppose that $H\overline{H} = 1$ and the first of the above alternatives holds. Combining the facts that $cH = a$, $H\overline{H} = 1$, and $c = \overline{a}\underline{a}$, we get $a\overline{a}^{[3]} = 1$, which is solved by $a = p_6/\overline{p}_6^{[3]}$. For c this solution a gives

$$c = \frac{\underline{p}_6\overline{p}_6}{\underline{\underline{p}}_6\overline{\overline{p}}_6},$$

while the h_n must satisfy $H = a/\underline{a}\overline{a}$.

Consider then the second alternative. By combining the known facts (assuming now that $h_1 h_2 \overline{h}_1 \overline{h}_2 = 1$) we get the equation $h_3 \overline{h}_3 \underline{\underline{h}}_3 \underline{h}_3 \overline{\overline{a}} = a$. The general solution of this is unknown, but it is solved for instance by $a = p_2$, $h_3 = p_4/\overline{\overline{p}}_4$.

Supposing that the h_n are period-1 we have $H \equiv \pm 1$ or $H = h_3$ in the alternative case. In the first case we get $\overline{a}\underline{a} = \pm a = c$. This equation has at least some solutions (see section 3.3.1). The alternative case gives $c^2 = -a\underline{a}$, which has a number of solutions (e.g. $c \equiv i$, $a = p_2/\overline{p}_2$).

Case $p = 2$, $q = 1$. Either $\overline{c}\overline{h}_1\overline{h}_2\underline{a} = \overline{a}c$ or $-\underline{c}a\overline{c} = \overline{\overline{a}}ch_1h_2\underline{\underline{a}}$, and the h_n satisfy one of 2a–2b.

Assuming that $H\overline{H} = 1$, we find from the first possible condition the equation $\underline{\underline{a}}\underline{a}\overline{c} = a\overline{a}\underline{c}$. This is solved for instance by $a = p_2\lambda^z$, $c = q_2\lambda^{2z}$. These imply that

$$\overline{H} = \frac{1}{H} = q_2/\overline{q}_2.$$

Thus, depending on which of the two possibilities the functions h_n satisfy, we find either of the following two equations:

$$(\overline{w}w - 1)(\underline{w}w - 1) = \frac{\overline{p}_2\lambda^{2z}(w - p_2)(w - 1/\overline{p}_2)}{w - q_2\lambda^z},$$

$$(\overline{w}w - 1)(\underline{w}w - 1) = \frac{\overline{p}_2\overline{q}_2\lambda^{2z}(w - h_1)(w - h_2)}{w - g_2\lambda^z}.$$

In the latter, h_1 and h_2 both solve (3.13).

The other possible contraint, again assuming $H\overline{H} = 1$ yields the equation $a\overline{a}\underline{c}\overline{\overline{c}} = \overline{\overline{a}}\overline{a}^{[3]}\underline{\underline{a}}\underline{a}$. This is solved at least by $a = c = p_3\lambda^z$, and we obtain more possible equations resembling the above two.

Restricting the h_n to be period-1 gives $H \equiv \pm 1$ so that $\overline{c}\underline{a} = \pm\overline{a}c$ or $\pm\underline{c}a\overline{c} = \overline{\overline{a}}c\underline{\underline{a}}$. With $H = +1$, the first of these has at least the solution $a = p_2\lambda^z$, $c = p_1\lambda^{2z}$, which gives

$$(w\overline{w} - 1)(w\underline{w} - 1) = \frac{p_1\lambda^{2z}(w - q_1)(w - 1/q_1)}{w - p_2\lambda^z}.$$

With $H = -1$ the first equation has at least the solution $a = p_2$, $c = p_1 e^{i\pi z}$, and then we have

$$(w\overline{w} - 1)(w\underline{w} - 1) = \frac{p_1 e^{i\pi z}(w^2 - 1)}{w - p_2}.$$

Case $p = 4$, $q = 0$. By Lemma 5.2, $cH = 1$. The h_n satisfy one of 4a–4e.

If $H\overline{H} = 1$ is true, we get $c\overline{c} = 1$. If the h_n are period-1, we have again $H \equiv \pm 1$, so that $c \equiv \pm 1$. Hence, we find either of the following equations:

$$(w\overline{w} - 1)(w\underline{w} - 1) = (w - p_1)(w - 1/p_1)(w - q_1)(w - 1/q_1),$$

$$(w\overline{w} - 1)(w\underline{w} - 1) = -(w - p_1)(w - 1/p_1)(w^2 - 1).$$

Case $p = 3$, $q = 0$. Now $cH = -1$ by Lemma 5.2. Either the h_n satisfy one of 3a–3c, or one of them, say h_3, satisfies $\overline{c}c\underline{\underline{h}}_3\overline{h}_3^{[3]} = 1$ while the others satisfy 2a or 2b.

The above equation for c and h_3 is solved at least by

$$c = \frac{p_2}{\overline{p}_2}\lambda^z, \quad h_1 = \frac{p_{10}}{\overline{p}_{10}^{[5]}}\lambda^{-z},$$

and if $H\overline{H} = 1$, we find $c\overline{c} = 1$, so that $\lambda = 1$.

If the h_n are assumed to be period-1 we have either $H \equiv \pm 1$, so that $c \equiv \pm 1$, or $\overline{c}ch_3^2 = 1$ and $ch_3 = -1$, which combine into $c = -1/h_3$. Hence we have the following examples:

$$\begin{aligned}(w\overline{w} - 1)(w\underline{w} - 1) &= \pm(w \pm 1)(w - 1/p_1)(w - 1/p_1),\\ (w\overline{w} - 1)(w\underline{w} - 1) &= -\frac{1}{p_1}(w - p_1)(w - q_1)(w - 1/q_1).\end{aligned}$$

Case $p = 2$, $q = 0$. Either the h_n satisfy 2a or 2b, or the relation $\overline{\overline{h}}_1c\underline{\underline{h}}_2 \equiv 1$ holds.

If the h_n are period-1, this latter option implies $ch_1h_2 \equiv 1$, while in the first case we find either of the following:

$$\begin{aligned}(w\overline{w} - 1)(w\underline{w} - 1) &= c(w - p_1)(w - 1/p_1),\\ (w\overline{w} - 1)(w\underline{w} - 1) &= c(w^2 - 1).\end{aligned}$$

Here we have not attempted to solve the coefficient c any further.

Case $p = 1$, $q = 0$. Now there are three possibilities: $h\overline{h} = 1$, or $c\underline{\underline{h}} = \overline{c}\overline{h}$, or $\overline{h}^{[3]}\underline{h}_{[3]}\overline{c}\underline{c} = -ch \equiv 1$.

The second one is solved at least by $c = p_1\lambda^z$, $h = p_3\lambda^{-z/3}$. In the third case the right hand side of (3.8) becomes $cw + 1$, where c satisfies $\overline{c}\underline{c} = \overline{c}^{[3]}\underline{c}_{[3]}$. This is solved for instance by $c = p_4\lambda^z u_{16}^{z^2}$.

If h is period-1, the first possibility implies that $h \equiv \pm 1$, the second that c is period-1, as well, and the third that $c = -1/h$. These put the right hand side of (3.8) into $c(w \pm 1)$, $p_1(w - q_1)$, or $p_1w + 1$, respectively.

Case $p = q = 0$. This case is just

$$(\overline{w}w - 1)(\underline{w}w - 1) = c. \tag{5.7}$$

6. Discussion

6.1. **Summary and review of the results.** We have shown that the existence of just one admissible meromorphic solution of hyper-order less than one is a sufficient condition to single out a relatively short list of difference equations from the large family of equations (3.4), assuming that the solution does not satisfy a difference Riccati equation. The lists of equations include the difference Painlevé equations III and V, as well as a number of difference equations closely resembling them. Many of these additional equations are known to be of Painlevé type, and integrable at least in the autonomous case.

We remark that the results we are able to rigorously prove are somewhat distinct from the actual hypothesis. The hypothesis, originally presented in [1], is that *sufficiently many* solutions of slow growth implies integrability, whether these solutions are admissible or not. In our results, just *one* solution of slow growth is needed, but it must be an admissible one.

The results as "Painlevé type" equations. While the differential Painlevé equations have a unique canonical form up to a Möbius transformation, the situation is much more complicated in the discrete setting. One criterion (but not alone a sufficient one) for a difference equation to be a difference Painlevé equation or of Painlevé type is the existence of a continuous limit to one of the differential Painlevé equations. There are, however, several equations that have such a limit to, say, the differential Painlevé I equation, and all of these very different looking equations can sometimes be referred to as "difference Painlevé I". Moreover, some of the difference Painlevé equation have continuous limits to more than one differential Painlevé equation, so that the choice of the limiting process determines the resulting differential equation.

Although we have identified equation (3.17a) as "the difference Painlevé III equation", it contains as its special cases equations that should rather be called difference Painlevé I or II, because they possess a continuous limit to the differential Painlevé I and II equations. Similarly, "the difference Painlevé V equation" (3.20a) reduces in some cases to forms that ought to be called difference Painlevé I, II, or IV.

The idea taken from the theory of differential equations, that there should be a set of canonical equations, "the discrete Painlevé equations I–VI" for instance, does not seem to be valid, and this "numbering" of the discrete Painlevé equations is in that sense misleading. A satisfactory complete theory of these equations and their relations does not exist at present, although attempts towards such a theory – for instance the algebro-geometric approach of Sakai [43] – exist.

Many of the equations we have discovered can undoubtedly be found, in some form or another, scattered in the vast literature on the subject of integrable discrete systems. Degenerate cases of the difference Painlevé III and V were studied and shown to be of Painlevé type in [37]. These include many of the same equations that we have found. Of course, there is a "change of setting" involved, since we consider analytic solutions while the objects of study in [37] are sequences. We also gave some examples in Section 3.3.2, but this still leaves many of the equations we have found unidentified in the sense that we do not know whether they are of Painlevé type or whether they are integrable (or indeed: whether they have any meromorphic solutions whatsoever).

It should also be remarked that in the cases when w was found to satisfy a Riccati difference equations, we did not proceed to investigate the forms of equations (3.5) or (3.8) further. This is in fact only natural, since several equations that are considered non-integrable have special Riccati solutions. For example, if w_0 is a solution to the

Riccati difference equation

$$\overline{w} = \frac{aw + b}{w + c},$$

then w_0 also satisfies

$$\underline{w} = \frac{-\underline{c}w + \underline{b}}{w - \underline{a}},$$

and by combining these we obtain

$$\overline{w}\underline{w} = \frac{(-\underline{c}w + \underline{b})(aw + b)}{(w - \underline{a})(w + c)},$$

which is generically not of Painlevé type, even though it possesses the special solution w_0.

Integrability and some examples. The difference Painlevé III and V equations and many of their degenerate cases are widely considered to be integrable. A difference Riccati equation (3.6) has a large class of meromorphic solutions, but the existence of finite-order solutions is so far known only in the autonomous case.

Many, but not all of the equations we have found fall into the QRT family in the autonomous case. Those that do are generically solvable in terms of elliptic functions, but again the non-autonomous case is still very much unknown. For example, the autonomous version of equation (3.17b),

$$\overline{w}\underline{w} = \frac{w^2 + c}{w^2 - 1}, \tag{6.1}$$

is not in the QRT family. To see this, one can write equation (3.1) into the form of (6.1), which gives a system of equations for the numbers $\alpha_i, \ldots, \zeta_i$ in the matrices that define the functions f_j (see page 8). It is not difficult to see that this system has no solutions.

For the most simple non-autonomous equations it is possible to find some concrete examples. Consider equation (3.7d). It is easy to find several examples for $q = 2$: The easiest case, $\overline{w}\underline{w} = w^2$, we have discussed in Section 3.3.1. With a non-constant h, some examples are found through the linearization

$$\overline{u} = \overline{h}u, \quad \overline{w} = uw,$$

for instance $(w, u, h) = (e^{z^3}, e^{3z^2+3z+1}, e^{6z})$, or $(w, u, h) = (\Gamma, z, z/(z-1))$, where Γ denotes the Gamma function. Such examples have a tiered relation in the growth in general, especially if the functions are entire.

For $q = -2$ and $q = 0$ we can find examples using tangent. Namely, if $w(z) = \tan(\pi z/2)$, then $\overline{w}\underline{w} = 1/w^2$, and if $w(z) = \tan(\pi z/4)$, then $\overline{w}\underline{w} = -1$. Examples for various nonconstant functions h can of course be generated by adding a small coefficient to w. The case $q = 1$ seems particularly problematic, as can be seen from the simplest possible case of equation (3.14).

The simplest case of equation (5.7), when $c \equiv 1$, can be written as

$$w = 1/\underline{w} + 1/\overline{w},$$

which is solved at least by $w = \sqrt{2}p_2/\overline{p}_2$. This, however, is also a solution to $\overline{w}\underline{w} = 2/w^2$. Examples that solve some equation of the form (5.7) but do not solve any equation of the form (3.5) appear to be hard to find.

These trivial examples perhaps serve to show how difficult it is to find concrete examples for the more complicated equations.

6.2. **Open questions.** It is known that the differential Painlevé V equation is in close relationship with the differential Painlevé III. More precisely, with a particular choice of parameters, a differential Painlevé V equation reduces to a differential Painlevé III for another function; see [12]. As remarked in [37], implementing the same condition to the discrete Painlevé V one should expect the equation

$$(\overline{w}w - 1)(w\underline{w} - 1) = \frac{\gamma(w^2+1)^2}{\alpha w^2 + \beta w + \gamma}$$

to be equivalent to some discrete Painlevé III equation (after some transformation on w). The question whether such a relation exists in the discrete case remains open. It is of course also possible that the relationship in the discrete case is not quite so directly an analogy to the differential case.

We used practically identically methods, even at times the same reasoning *verbatim*, in the treatments that yielded the difference Painlevé III and V equations. This alone suggests strongly that there is some connection between these two equations.

Another large open question, which has already been introduced above, is the lack of generic methods to solve the non-autonomous equations, and even the lack of concrete example solutions for such equations.

Whereas the list of equations in Theorem 3.2 contains all known integrable equations in the class (3.2) and apparently no non-integrable ones, we have not been able to identify all of the equations arising from the study of the families of the difference Painlevé III and V in the same manner, or even to list them in a similar closed form. Our results could possibly be further restricted with some additional reasoning. The difference Painlevé I and II equations also seem to be more widely studied in the existing literature.

6.3. **Alternative approaches.** There are several ways to study the integrability of difference equations in various settings. A few of these methods, mainly those that are widely used and closely connected to the criterion used in this thesis, are briefly discussed in what follows.

On the method of singularity confinement. Our approach to the question of integrability of difference equations – to require the existence of meromorphic solutions with sufficiently small growth – is closely related to the singularity confinement (SC) method of [10], which was used in the discovery of the discrete versions of the Painlevé equations in [38].

The simple idea of singularity confinement is to demand that whenever an iterate of the equation becomes infinite, the following iterates must remain finite and the data of the initial conditions must be preserved when passing the singularity.

Moreover, to deal with examples such as $x_{n+1} = x_n^2/(x_n^2 - 1)$, it is required that backward iteration is well defined without proliferation of preimages.

To allow equations such as

$$\overline{w}\underline{w} = \frac{w^2 + k^2}{w^2 + 1}, \tag{6.2}$$

to pass the test, only singularities that have a definite starting point need to be confined. Equation (6.2), which is a special case of the difference Painlevé III equation, has a one-parameter family of elliptic solutions $w = i/\mathrm{sn}(z + c, k)$, where $c \in \mathbb{C}$ and the modulus k is such that the Jacobi sn function $\mathrm{sn}(z, k)$ has a real period $2K = 2$. By the properties of sn, the poles of such a solution w appear in infinite sequences.

The SC test is relatively easy to use and has been applied widely in the study of integrable difference equations. There exist, however, well known examples showing that SC is not sufficient for integrability. Hietarinta and Viallet found in [23] an example that passes the SC test but is still chaotic. They proposed the additional requirement that the equation should have zero *algebraic entropy*

$$\lim_{n\to\infty} \frac{\log d_n}{n},$$

where d_n is the degree of the iterates of the equation (considered as a rational function of the initial conditions).

The example of Hietarinta and Viallet is the equation

$$y_{n+1} + y_{n-1} = y_n + \frac{a}{y_n^2},$$

which in our setting would read

$$\overline{y} + \underline{y} = y + \frac{a}{y^2}.$$

Using the same methods as in Chapters 4 and 5, it can be shown that any meromorphic solution y has hyper-order at least one. This, of course, is a known fact, but one that can indeed be obtained also using our method. We will omit the details here; see [16].

Diophantine integrability. Let κ be some number field, i.e. a finite extension of the field of rational numbers. The height $H(x)$ of an element $x \in \kappa$ measures the complexity of x. In the simplest case, when $\kappa = \mathbb{Q}$, the height of a nonzero rational number is $H(p/q) = \max\{|p|, |q|\}$, when p and q have no common factors. We skip the more complicated definition of the height in a general number field; see for instance [3] for further details.

The notion of height in a number field is in many ways analogous to the Nevanlinna characteristic of a meromorphic function, as observed by Osgood [34, 35] and Vojta [48]. Many, but not all of the results in Nevanlinna theory have a more or less direct analogue in algebraic number theory. Vojta constructed a "dictionary" for translating the concepts of one theory to the other, with the aim that a statement in Nevanlinna theory about the characteristic of a meromorphic function should correspond to a statement in Diophantine approximation about an infinite set of elements in a number field (the radii r correspond to individual elements).

The dictionary is not complete, however, as there is no clear analogue for the derivative of a meromorphic function, which plays an important role in Nevanlinna theory. Finding such an analogue would quite likely solve many existing problems and conjectures in number theory. It might also be possible to build a similar analogy by using the difference approach of Theorem 3.5 to Nevanlinna theory, instead of the usual lemma on the logarithmic derivative. For the difference operator $\Delta f(z) := f(z+1) - f(z)$ there clearly is a simple discrete analogue.

For more details on the relationship of Diophantine approximation and Nevanlinna theory, see for instance [4, 3]. The correspondences between the theories are studied in more depth and width in [41].

The criterion of integrability that has been applied in this thesis can, at least under certain assumptions, be translated using Vojta's dictionary. In [13], the following definition was suggested: a polynomial discrete equation for y_n is *Diophantine integrable* if the logarithmic height of the iterates $h(y_n) = \log H(y_n)$ grows no faster than a polynomial in n.

Let w be a finite-order meromorphic solution to a given difference equation. Restricting the independent variable of w to integers we find a discrete equation for y_n, say. Assume that after this restriction all the iterates y_n are in some number field κ (for suitable initial values chosen in κ). Then the fact that w was of finite order would, by Vojta's dictionary, correspond to $h(y_n)$ growing no faster than a polynomial in n. That these statements would in fact be equivalent has so far been proven only for certain first-order equations, see [13].

Diophantine integrability is related to the zero algebraic entropy condition of Hietarinta and Viallet. If

$$R = \frac{a_p x^p + a_{p-1}x^{p-1} + \dots a_1 x + a_0}{b_q x^q + b_{q-1}x^{q-1} + \dots b_1 x + b_0}$$

is an irreducible rational function of x and $\deg R = \max\{p, q\} =: d$, then the logarithmic height satisfies

$$|h(R) - dh(x)| < \log C,$$

where C is a polynomial in $H(a_i)$ and $H(b_j)$ [13] (this can be seen as an analogue of Theorem 3.6). In the Diophantine integrability approach, the height growth is considered directly, and the required condition is quicker to check numerically even for a large number of iterates.

Several results of Nevanlinna theory can be directly translated to the study of Diophantine integrability. See [13] for more details.

Other settings. The general principle that integrability seems to correlate with the slow growth of certain characteristics was expressed by Veselov [47], who showed that certain types of first integrals do not exist for equations where the degree of the iterates grows exponentially. Roberts and Vivaldi [40] studied orbit dynamics over finite fields, and found markedly different orbit statistics for integrable mappings than for nonintegrable ones. The approach of estimating the growth of solutions has also been applied in the study of ultradiscrete equations, and a tropical version of Nevanlinna theory has been developed for this purpose, see [20].

References

1. M. J. Ablowitz, R. G. Halburd, and B. Herbst, *On the extension of the Painlevé property to difference equations*, Nonlinearity **13** (2000), 889–905.
2. D. C. Barnett, R. G. Halburd, R. J. Korhonen, and W. Morgan, *Nevanlinna theory for the q-difference operator and meromorphic solutions of q-difference equations*, Proc. Roy. Soc. Edinburgh, Sect. A. **137** (2007), 457–474.
3. E. Bombieri and W. Gubler, *Heights in Diophantine geometry*, New Mathematical Monographs, vol. 4, Cambridge University Press, 2006.
4. W. Cherry and Z. Ye, *Nevanlinna's theory of value distribution*, Springer-Verlag, Berlin, 2001.
5. Y. M. Chiang and S. J. Feng, *On the Nevanlinna characteristic of $f(z+\eta)$ and difference equations in the complex plane*, Ramanujan J. **16** (2008), 105–129.
6. A. S. Fokas, A. R. Its, and A. V. Kitaev, *Discrete Painlevé equations and their appearance in quantum gravity*, Comm. Math. Phys. **142** (1991), 313–344.
7. B. Grammaticos, R. G. Halburd, A. Ramani, and C.-M. Viallet, *How to detect the integrability of discrete systems*, preprint (available at: `http://www.newton.ac.uk/preprints/NI09060.pdf`), 2009.
8. B. Grammaticos, Y. Kosmann-Schwarzbach, and T. Tamizhmani (eds.), *Discrete integrable systems*, Lecture Notes in Physics, vol. 644, Berlin, Springer-Verlag, 2004.
9. B. Grammaticos, F. W. Nijhoff, V. Papageorgiou, A. Ramani, and J. Satsuma, *Linearization and solutions of the discrete Painlevé III equation*, Phys. Lett. A **185** (1994), 446–452.
10. B. Grammaticos, A. Ramani, and V. Papageorgiou, *Do integrable mappings have the Painlevé property?*, Phys. Rev. Lett. **67** (1991), 1825–1828.
11. B. Grammaticos, T. Tamizhmani, A. Ramani, and K.M. Tamizhmani, *Growth and integrability in discrete systems*, J. Phys. A: Math. Gen. **34** (2001), 3811–3821.
12. V. I. Gromak, I. Laine, and S. Shimomura, *Painlevé differential equations in the complex plane*, Walter de Gruyter, Berlin, 2002.
13. R. G. Halburd, *Diophantine integrability*, J. Phys. A: Math. Gen. **38** (2005), L1–L7.
14. R. G. Halburd and R. J. Korhonen, *Difference analogue of the lemma on the logarithmic derivative with applications to difference equations*, J. Math. Anal. Appl. **314** (2006), 477–487.
15. ______, *Existence of finite-order meromorphic solutions as a detector of integrability in difference equations*, Physica D **218** (2006), 191–203.
16. ______, *Finite-order meromorphic solutions and the discrete Painlevé equations*, Proc. Lond. Math. Soc. **3** (2006), 1–32.
17. ______, *Nevanlinna theory for the difference operator*, Ann. Acad. Sci. Fenn. Math. **31** (2006), 463–478.
18. ______, *Meromorphic solutions of difference equations, integrability and the discrete Painlevé equations*, J. Phys. A: Math. Theor. **40** (2007), R1–R38.
19. R. G. Halburd, R. J. Korhonen, and K. Tohge, *Holomorphic curves with shift-invariant hyperplane preimages*, preprint (arXiv:0903.3236), 2009.
20. R. G. Halburd and N. J. Southall, *Tropical Nevanlinna theory and ultradiscrete equations*, Int. Math. Res. Not. **2009** (2009), 887–911.

21. W. K. HAYMAN, *Meromorphic functions*, Clarendon Press, Oxford, 1964.
22. J. HIETARINTA, private communication.
23. J. HIETARINTA AND C. VIALLET, *Singularity confinement and chaos in discrete systems*, Phys. Rev. Lett. **81** (1998), 325–328.
24. M. JIMBO, H. SAKAI, A. RAMANI, AND B. GRAMMATICOS, *Bilinear structure and Schlesinger transforms of the q-P_{III} and q-P_{IV} equations*, Phys. Lett. A **217** (1996), 111–118.
25. K. KAJIWARA, Y. OHTA, AND J. SATSUMA, *Casorati determinant solutions for the discrete Painlevé-III equation*, J. Math. Phys. **36** (1995), 4162–4174.
26. K. KATAJAMÄKI, *Algebroid solutions of binomial and linear differential equations*, Ann. Acad. Sci. Fenn. Math. Diss. **90** (1993), 48pp.
27. K. KIMURA, H. YAHAGI, R. HIROTA, A. RAMANI, B. GRAMMATICOS, AND Y. OHTA, *A new class of integrable discrete systems*, J. Phys. A **35** (2002), 9205–9212.
28. R. J. KORHONEN, *A new Clunie type theorem for difference polynomials*, to appear in J. Difference Equ. Appl. (2009), 12pp.
29. I. LAINE, *Nevanlinna theory and complex differential equations*, Walter de Gruyter, Berlin/New York, 1993.
30. I. LAINE, J. RIEPPO, AND H. SILVENNOINEN, *Remarks on complex difference equations*, Comput. Methods Funct. Theory **5** (2005), 77–88.
31. I. LAINE AND C.-C. YANG, *Clunie theorems for difference and q-difference polynomials*, J. Lond. Math. Soc. **76** (2007), 556–566.
32. L. M. MILNE-THOMSON, *The calculus of finite differences*, AMS Chelsea Pub., Providence, R.I., 2000.
33. A. Z. MOHON'KO, *The Nevanlinna characteristics of certain meromorphic functions*, Teor. Funktsii Funktsional. Anal. i Prilozhen **14** (1971), 83–87, (Russian).
34. C. F. OSGOOD, *A number theoretic differential equations approach to generalizing Nevanlinna theory*, Indian J. Math. **23** (1981), 1–15.
35. ———, *Sometimes effective Thue-Siegel Roth Schmidt-Nevanlinna bounds, or better*, J. Number Theory **21** (1984), 347–389.
36. V. G. PAPAGEORGIOU, F. W. NIJHOFF, B. GRAMMATICOS, AND A. RAMANI, *Isomonodromic deformation problems for discrete analogues of Painlevé equations*, Phys. Lett. A **164** (1992), 57–64.
37. A. RAMANI AND B. GRAMMATICOS, *Discrete Painlevé equations: coalescences, limits and degeneracies*, Physica A **228** (1996), 160–171.
38. A. RAMANI, B. GRAMMATICOS, AND J. HIETARINTA, *Discrete versions of the Painlevé equations*, Phys. Rev. Lett. **67** (1991), 1829–1832.
39. A. RAMANI, B. GRAMMATICOS, AND J. SATSUMA, *Bilinear discrete Painlevé equations*, J. Phys. A **28** (1995), no. 16, 4655–4665.
40. J. A. G. ROBERTS AND F. VIVALDI, *Arithmetical method to detect integrability in maps*, Phys. Rev. Lett. **90** (2003), 034102.
41. M. RU, *Nevanlinna theory and its relation to Diophantine approximation*, World Scientific River Edge, Singapore, 2001.
42. S. N. M. RUIJSENAARS, *First order analytic difference equations and integrable quantum systems*, J. Math. Phys. **38** (1997), no. 2, 1069–1146.
43. H. SAKAI, *Rational surfaces associated with affine root systems and geometry of the Painlevé equations*, Comm. Math. Phys. **220** (2001), 165–229.
44. A. SAKKA AND U. MUĞAN, *Bäcklund transformations for discrete Painlevé equations: Discrete P_{II}–P_V*, Chaos Solitons Fractals **27** (2006), 387–294.
45. K. M. TAMIZHMANI, A. RAMANI, B. GRAMMATICOS, AND Y. OHTA, *A study of the discrete P_V equation: Miura transformations and particular solutions*, Lett. Math. Phys. **38** (1996), 289–296.

46. G. Valiron, *Sur la dérivée des fonctions algébroïdes*, Bull. Soc. Math. France **59** (1931), 17–39.
47. A. P. Veselov, *Growth and integrability in the dynamics of mappings*, Commun. Math. Phys. **145** (1992), 181–193.
48. P. Vojta, *Diophantine approximations and value distribution theory*, Lecture Notes in Mathematics, vol. 1239, Springer, Berlin, 1987.
49. J. M. Whittaker, *Interpolatory function theory*, Cambridge Tracts in Mathematics and Mathematical Physics, vol. 33, Stechert-Hafner, New York, 1964.
50. N. Yanagihara, *Meromorphic solutions of some difference equations*, Funkcialaj Ekvacioj **23** (1980), 309–326.
51. ———, *Meromorphic solutions of some difference equations and the value distribution theory in half-strips*, Value distribution theory and its applications (New York, 1983), Contemp. Math., vol. 25, Amer. Math. Soc., Providence, RI, 1983, pp. 245–253.

Department of Mathematics, University of Eastern Finland, P.O. Box 111, FI-80101, Finland

E-mail address: onni.ronkainen@uef.fi